Bruni?
October 13, 97
Ithaca NY

D0463382

Many important plant crop products are produced in non-photosynthetic tissues such as seeds, roots and tubers, in which the provision of energy for biosynthetic processes is different from that found in green cells. In this book, leading researchers in the field present a fine overview of the latest thinking on the organization of intermediary metabolism in such organs, and the development of some of the cellular compartments responsible for the synthesis of important crop products. Graduate students as well as researchers in the field of plant biochemistry will find this book of great interest.

SOCIETY FOR EXPERIMENTAL BIOLOGY
SEMINAR SERIES: 42

COMPARTMENTATION OF PLANT METABOLISM IN
NON-PHOTOSYNTHETIC TISSUES

SOCIETY FOR EXPERIMENTAL BIOLOGY SEMINAR SERIES

A series of multi-author volumes developed from seminars held by the Society for Experimental Biology. Each volume serves not only as an introductory review of a specific topic, but also introduces the reader to experimental evidence to support the theories and principles discussed, and points the way to new research.

1. Effects of air pollution on plants. *Edited by T. A. Mansfield*
2. Effects of pollutants on aquatic organisms. *Edited by A.P.M. Lockwood*
3. Analytical and quantitative methods. *Edited by J. A. Meek and H. Y. Elder*
4. Isolation of plant growth substances. *Edited by J. R. Hillman*
5. Aspects of animal movement. *Edited by H. Y. Elder and E. R. Trueman*
6. Neurones without impulses: their significance for vertebrate and invertebrate systems. *Edited by A. Roberts and B.M.H. Bush*
7. Development and specialisation of skeletal muscle. *Edited by D. F. Goldspink*
8. Stomatal physiology. *Edited by P.G. Jarvis and T.A. Mansfield*
9. Brain mechanisms of behaviour in lower vertebrates. *Edited by P. R. Laming*
10. The cell cycle. *Edited by P.C.L. John*
11. Effects of disease on the physiology of the growing plant. *Edited by P.G. Ayres*
12. Biology of the chemotactic response. *Edited by J.M. Lackie and P.C. Williamson*
13. Animal migration. *Edited by D. J. Aidley*
14. Biological timekeeping. *Edited by J. Brady*
15. The nucleolus. *Edited by E.G. Jordan and C.A. Cullis*
16. Gills. *Edited by D.F. Houlihan, J.C. Rankin and T.J. Shuttleworth*
17. Cellular acclimatisation to environmental change. *Edited by A.R. Cossins and P. Sheterline*
18. Plant biotechnology. *Edited by S. H. Mantell and H. Smith*
19. Storage carbohydrates in vascular plants. *Edited by D.H. Lewis*
20. The physiology and biochemistry of plant respiration. *Edited by J.M. Palmer*
21. Chloroplast biogenesis. *Edited by R.J. Ellis*
22. Instrumentation for environmental physiology. *Edited by B. Marshall and F. I. Woodwood*
23. The biosynthesis and metabolism of plant hormones. *Edited by A. Crozier and J.R. Hillman*
24. Coordination of motor behaviour. *Edited by B.M.H. Bush and F. Clarac*
25. Cell ageing and cell death. *Edited by I. Davies and D.C. Sigee*
26. The cell division cycle in plants. *Edited by J. A. Bryant and D. Francis*
27. Control of leaf growth. *Edited by N.R. Baker, W.J. Davies and C. Ong*
28. Biochemistry of plant cell walls. *Edited by C. T. Brett and J. R. Hillman*
29. Immunology in plant science. *Edited by T. L. Wang*
30. Root development and function. *Edited by P.J. Gregory, J.V. Lake and D.A. Rose*
31. Plant canopies: their growth, form and function. *Edited by G. Russell, B. Marshall and P.G. Jarvis*
32. Developmental mutants in higher plants. *Edited by H. Thomas and D. Grierson*
33. Neurohormones in invertebrates. *Edited by M. Thorndyke and G. Goldsworthy*
34. Acid toxicity and aquatic animals. *Edited by R. Morris, E.W. Taylor, D.J.A. Brown and J.A. Brown*
35. Division and segregation of organelles. *Edited by S.A. Boffey and D. Lloyd*
37. Techniques in comparative respiratory physiology: An experimental approach. *Edited by C.R. Bridges and P.J. Butler*
38. Herbicides and plant metabolism. *Edited by A.D. Dodge*
39. Plants under stress. *Edited by H.G. Jones, T.J. Flowers and M.B. Jones*
40. *In situ* hybridisation: application to developmental biology and medicine. *Edited by N. Harris and D.G. Wilkinson*
41. Physiological strategies for gas exchange and metabolism. *Edited by A.J. Woakes, M.K. Grieshaber and C.R. Bridges*
42. Compartmentation of plant metabolism in non-photosynthetic tissues. *Edited by M.J. Emes*
43. Plant Growth: interactions with nutrition and environment. *Edited by J.R. Porter and D.W. Lawlor*

COMPARTMENTATION OF PLANT METABOLISM IN NON-PHOTOSYNTHETIC TISSUES

Edited by

M.J. Emes

Senior Lecturer, Department of Cell and Structural Biology
University of Manchester

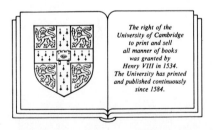

The right of the
University of Cambridge
to print and sell
all manner of books
was granted by
Henry VIII in 1534.
The University has printed
and published continuously
since 1584.

CAMBRIDGE UNIVERSITY PRESS

Cambridge

New York   Port Chester   Melbourne   Sydney

Published by the Press Syndicate of the University of Cambridge
The Pitt Building, Trumpington Street, Cambridge CB2 1RP
40 West 20th Street, New York, NY 10011-4211, USA
10 Stamford Road, Oakleigh, Melbourne 3166, Australia

© Cambridge University Press 1991

First published 1991

Printed in Great Britain at the University Press, Cambridge

British Library Cataloguing in publication data
Compartmentation of plant metabolism in non-photosynthetic tissues.
 1. Plants. Cells. Metabolism
 I, Emes, M. J.
 581.8761

Library of Congress cataloguing in publication data
Compartmentation of plant metabolism in non-photosynthetic tissues/
 edited by M. J. Emes.
  p. cm. – (Society for Experimental Biology seminar series; 42)
 Includes index.
ISBN 0-521-36132-X
 1. Plants – Metabolism – Congresses. 2. Plant cell compartmentation –
Congresses. 3. Plant cells and tissues – Congresses. I. Emes, M. J. II. Title:
Non-photosynthetic tissues. III. Series: Seminar series (Society for Experimental
Biology (Great Britain)); 42.
QK881.C66 1991
581.1′33 – dc20 90-44516 CIP

ISBN 0 521 36132 X hardback

wv

CONTENTS

Contributors   ix
Preface   xiii

Metabolic compartmentation in plant cells   1
  H. BEEVERS

Fatty acid and lipid biosynthesis   23
  J.L. HARWOOD

**Structure, synthesis and degradation of oil bodies in
maize**   43
  A.H.C. HUANG, R. QU, Y.K. LAI,
  C. RATNAYAKE, K.L. CHAN,
  G.W. KUROKI, K.C. OO AND Y.Z. CAO

**Biogenesis of protein bodies and glyoxysomes in *Ricinus
communis* seeds**   59
  J.M. LORD, C. HALPIN, M.J. CONDER
  AND S.D. IRWIN

Isozymes and compartmentation in leucoplasts   77
  D.T. DENNIS, S. BLAKELEY
  AND S. CARLISLE

**Interconversion of C-6 and C-3 sugar phosphates in
non-photosynthetic cells of plants**   95
  T. AP REES, T.G. ENTWISTLE
  AND J.E. DANCER

**The pathway and compartmentation of starch synthesis
in developing wheat grain**   111
  P.L. KEELING

Autophagy triggered by sucrose deprivation in sycamore
(*Acer pseudoplatanus*) cells 127
    R. DOUCE, R. BLIGNY, D. BROWN,
    A.-J. DORNE, P. GENIX AND C. ROBY

Integration and compartmentation of carbon and
nitrogen metabolism in roots 147
    M.J. EMES AND C.G. BOWSHER

Control of the rate of respiration in roots:
compartmentation, demand and the supply of substrate 167
    J.F. FARRAR AND J.H.H. WILLIAMS

Control of the rate of respiration in shoots: light,
calcium and plant growth regulators 189
    A.R. WELLBURN AND J.H. OWEN

Index 199

CONTRIBUTORS

ap Rees, T.
School of Botany, University of Cambridge, Cambridge
CB2 3EA, UK.

Beevers, H.
Dept. of Biology, University of California, Santa Cruz,
CA 95064, USA.

Blakeley, S.
Queen's University, Ontario K7L 3N6, Canada.

Bligny, R.
Dept. de Recherche Fondamentale, Centre d'Etudes
Nucléaires et Université Joseph Fourier, Grenoble-cédex
85X F38041, France.

Bowsher, C.G.
Dept. of Molecular Biology and Genetics, University of
Guelph, Ontario N1G 2W1, Canada.

Brown, D.
Dept. de Recherche Fondamentale, Centre d'Etudes
Nucléaires et Université Joseph Fourier, Grenoble-cédex
85X F38041, France.

Cao, Y.Z.
Dept. of Botany and Plant Sciences, University of California,
Riverside, CA 92521, USA.

Carlisle, S.
Queen's University, Ontario K7L 3N6, Canada.

Chan, K.L.
Dept. of Botany and Plant Sciences, University of California,
CA 92521, USA.

Conder, M.J.
Dept. of Biological Sciences, University of Warwick, Coventry
CV5 7AL, UK.

Dancer, J.E.

School of Botany, University of Cambridge, Cambridge
CB2 3EA, UK.
Dennis, D.T.
Queen's University, Ontario K7L 3N6, Canada.
Dorne, A.-J.
Dept. de Recherche Fondamentale, Centre d'Etudes
Nucléaires et Université Joseph Fourier, Grenoble-cédex
85X F38041, France.
Douce, R.
Dept. de Recherche Fondamentale, Centre d'Etudes
Nucléaires et Université Joseph Fourier, Grenoble-cédex,
85X F38041, France.
Emes, M.J.
Plant Metabolism Research Unit, University of Manchester,
Manchester M13 9PL, UK.
Entwistle, T.G.
School of Botany, University of Cambridge, Cambridge
CB2 3EA, UK.
Farrar, J.F.
School of Biological Sciences, University College of North
Wales, Gwynedd LL57 2UW, UK.
Genix, P.
Dept. de Recherche Fondamentale, Centre d'Etudes
Nucléaires et Université Joseph Fourier, Grenoble-cédex
85X F38041, France.
Halpin, C.
Dept. of Biological Sciences, University of Warwick, Coventry
CV5 7AL, UK.
Harwood, J.L.
Dept. of Biochemistry, University of Wales, College of
Cardiff, Cardiff CF1 1ST, UK.
Huang, A.H.C.
Dept. of Botany and Plant Sciences, University of California,
CA 92521, USA.
Irwin, S.D.
Dept. of Biological Sciences, University of Warwick, Coventry
CV5 7AL, UK.
Keeling, P.L.
Garst Seed Company, P.O. Box 500, Slater, IA 50244, USA.
Kuroki, G.W.
Dept. of Botany and Plant Sciences, University of California,
CA 92521, USA.

Lai, Y.K.
Dept. of Botany and Plant Sciences, University of California,
CA 92521, USA.
Lord, J.M.
Dept. of Biological Sciences, University of Warwick,
Coventry, CV5 7AL, UK.
Oo, K.C.
Dept. of Botany and Plant Sciences, University of California,
CA 92521, USA.
Owen, J.H.
Eton College, Windsor, Berkshire SL4 6DW, UK.
Qu, R.
Dept. of Botany and Plant Sciences, University of California,
Riverside, CA 92521, USA.
Ratnayake, C.
Dept. of Botany and Plant Sciences, University of California,
CA 92521, USA.
Roby, C.
Dept. de Recherche Fondamentale, Centre d'Etudes
Nucléaires et Université Joseph Fourier, Grenoble-cédex
85X F38041, France.
Wellburn, A.R.
Inst. of Environmental and Biological Sciences, University of
Lancaster, Lancaster LA1 4YQ, UK.
Williams, J.H.H.
School of Biological Sciences, University College of North
Wales, Gwynedd LL57 2UW, UK.

PREFACE

A number of books have been produced in recent years which aim to take account of the fact that the higher plant cell is a highly compartmentalized system. Consideration of the sub cellular site of a particular activity inevitably tends to be dominated by work carried out on leaf cells. Very often it turns out that the chloroplast is a major site of some important (biosynthetic) pathway further concentrating attention on the photosynthetic cells of a plant. Unfortunately it is often assumed that intracellular compartmentation and regulation of metabolism away from green cells follows the same line of that in the leaf. However, a brief consideration of the roles of non-photosynthetic tissues indicates that separate consideration needs to be given to the compartmentation of their activities. For example, non-chlorophyllous plastids are capable of a wide range of biosynthetic processes yet their requirements for anabolism such as ATP, reducing power and carbon skeletons would all have to be imported, unlike the case of the chloroplast where there is a ready supply of all three. Non-photosynthetic cells are a sink for sucrose produced in photosynthesis, and it would be expected that the control of sucrose metabolism, taking account of its compartmentalization, would be different from the source tissue, the leaf. Many of the most commercially important crop products such as oils, starch and protein are produced in non-photosynthetic developing seeds or tubers, and an understanding of the early biochemical changes which take place during germination is fundamental to an understanding of plant development. There are many other examples which one could choose and some of these will, I hope, be drawn out in this book.

Over the years a number of groups have been involved in the study of non-photosynthetic plant cell metabolism. In 1989, a meeting was held on the topic of metabolic compartmentation in non-photosynthetic plant cells organized by the Plant Metabolism Group of the SEB at Edinburgh (and generously supported by ICI Seeds and the Agricultural Genetics Company) in an attempt to draw together the various threads of this subject. The areas covered included compartmentation of metabolism in roots, endosperm, seeds, meristematic tissue and non-green cell cultures. This book is the product of that meeting. A number of the chapters of this book are concerned with non-green plastids reflecting the relatively greater amount of attention they have received. Other chapters are concerned with metabolism in lipid bodies, glyoxysomes, vacuoles, protein bodies and mitochondria. As well as studies of isolated organelles and enzymes, *in situ* experiments employing the power of NMR are described. It is a special pleasure to have an introductory overview from Professor Harry Beevers who has probably made the largest single contribution to this subject both directly and through his influence on others.

My thanks to each of the participants of that meeting for their contribution, time and patience and to Tom ap Rees for providing a valuable chapter even though, through force of circumstance, he could not attend the meeting itself. My sincere wish is that more plant biochemists will turn their attention to the study of metabolism in non-green plant cells. The technical problems often seem greater than those associated with leaf tissue but I would remind readers of the words of Alexander Solzhenitsyn

In exploring new ground, difficulties must be seen as buried treasure. In general, the greater they are the more valuable ... The most rewarding way of research is that of the greatest possible resistance from outside and least possible from within. Failure must be seen as the need for further effort and toughening of the will. And the more effort has been put in already, the more joy there is in meeting obstruction: it means that the pick has struck the iron casket which contains the treasure. And the value of overcoming increased difficulties is the greater because failure has made you proportionately to their size. (*The First Circle*.)

M.J. Emes
Manchester

H. BEEVERS

Metabolic compartmentation in plant cells

Introduction

Our predecessors, enlightened by Buchner's discovery, are said to have regarded the cell as a bag of enzymes. However apocryphal this view, the early biochemists did little to dispel it in their concentration on soluble enzymes and in the care they took to filter or clarify their extracts before assay. To later generations, the cytological complexity revealed by successively better light and electron microscopic techniques posed a challenge, the elucidation of the metabolic significance of the structures so revealed. In the past 50 years, indeed, much has been learned about structure: function relationships within cells; cytology has grown out of its preoccupation with nuclei and chromosomes and a field of cell biology has developed which acknowledges that there are other (more) interesting parts of cells. Procedures have been introduced for the isolation of many of these parts and metabolic roles have now firmly been assigned to most of the organelles, membranes and other structures seen dispersed in the cytosol in plant cells. The cell is now regarded as a unit within which individual reaction sequences are segregated from each other, yet working harmoniously as a whole through the transport of particular metabolites from one intracellular compartment to another.

In this overview of intracellular compartmentation of metabolism in non-photosynthetic plant tissues I will discuss the general consequences of compartmentation and how it can be demonstrated experimentally in living tissue before dealing with organelle isolation and other ways of determining which enzymes are confined to particular cell compartments.

The consequences of compartmentation

One important result of the compartmentation of enzymes of particular reaction sequences within organelles in plants (and all eukaryotic cells) is that the enzymes in the organelle matrix are much more concentrated than they would otherwise be. For example if the mitochondrial matrix occupies say two per cent of the total cytoplasmic volume, the concentration of the enzymes there would be roughly 50 times that which would be obtained if the same amount of enzyme was distributed evenly through the cytoplasm. As each type of organelle contains only a subset of the enzymes present in the cell, the chances of the product of an enzyme reaction encountering the subsequent enzyme in a sequence are correspondingly enhanced. Indeed within organelle matrices and the cytosol itself there are thought to be physical associations of enzymes which further increase the efficiency of overall catalysis (Srere, 1987) and in membranes the physical propinquity of sequential components, e.g. as in electron chains, is such that efficient channelling results.

In any living cell there is constant turnover of macromolecular components and during growth, rates of synthesis exceed those of breakdown. The segregation of enzymes concerned with synthesis in a compartment physically separate from that in which breakdown occurs allows the two opposing reactions to be independently regulated. In a more general sense individual reaction sequences in different organelles can be separately initiated, regulated or prevented through appropriate effectors or hormones targeted to particular organelles through receptors or transport systems. Correspondingly, coordinate regulation can be brought about through effectors or coenzymes moving between compartments.

It is now recognized that ion gradients across membranes can be used to drive transport or to generate ATP. In plant cells there are several distinct compartments across whose membranes such work may be done. Thus in addition to those in the mitochondria and plastids, proton or other ion gradients related to ATP production or consumption have also been recognized in plasma membrane, tonoplast and Golgi vesicles (Sze, 1985). Furthermore, it has been clear for many years that the major compartment of plant cells, the vacuole, is markedly more acidic (by up to 4 pH units) than the cytoplasm, and that the plastid matrix during photosynthesis is more alkaline than the cytosol. The variety of organelles present in plant cells allows a corresponding variety of different pH conditions (and ion concentrations generally) to be separately maintained in specific regions of the cell.

Compartmentation is also important in the transport of components generated in the cytosol through the plasmalemma. The role of the Golgi and vesicles derived from it in the secretion of enzymes and cell wall precursors is becoming clear (Mollenhauer & Morré, 1980; Akazawa & Hara-Nishimura, 1985).

Compartmentation also makes it possible to restrict compounds, including waste products and toxic materials, to a region of the cell out of contact with cytoplasmic components. The vacuole is the distinctive plant cell organelle that fulfills this and other roles (Marty, Branton & Leigh, 1980; Raven, 1987). The confinement of some water soluble pigments and phenolic compounds in vacuoles has long been known. More recently it has been recognized that the sequestration there of noxious constituents such as ricin (Harley & Beevers, 1982) and cyanogenic glycosides (Saunders & Conn, 1978) affords a vital separation from cytoplasmic enzymes. In some cell types the vacuole plays a major role in protein hydrolysis and many of the highly active hydrolases with acid pH optima are located there (Nishimura & Beevers, 1978b). In addition this is a compartment in which salts and metabolites such as sugars and acids can be temporarily stored.

The beneficial effects that accrue from compartmentation are not without metabolic cost. One obvious consequence is that particular metabolites must move efficiently from one compartment to another. Specific transport systems and shuttles are recognized; these imply that, in addition to the individual enzymes in a metabolic sequence, an organelle will contain particular proteins concerned with the implementation of transport across the membranes. The existence of such proteins has been deduced in most of the organelles isolated from plants; particular progress in their elucidation has been made in chloroplasts and mitochondria (Heldt & Flügge, 1987).

An additional complication arising from the fact of compartmentation is that during cell division and growth each of the classes of organelles must be replicated. As currently understood biogenesis of all organelles entails the insertion of proteins newly synthesized on cytoplasmic ribosomes into or through the membranes of developing organelles. To account for the accurate targeting of enzymes to the appropriate organelle, specific recognition sites on each class of organelle are invoked.

Compartmentation of metabolites in living cells

The division of cellular labour that we call metabolic compartmentation rests on the confinement of particular enzymes or sequences to distinctive organelles. An important corollary is that the metabolic intermediates in a pathway are consumed by the subsequent enzyme in that pathway; they may not dissociate from the enzyme complex and in any event are prevented from becoming equally distributed throughout the cell by the bounding membrane of the organelle. This measure of confinement of *metabolites* in a pathway to the organelle in which the enzymes are present leads to the conclusion that the total amount of a particular metabolite in the cell is likely to be comprised of more than one component, that in one organelle may be turning over rapidly and out of contact with other component(s) turning over more slowly or not at all. For example, the malate in the mitochondrial matrix may represent only a fraction of the total malate in the cell, and it will turn over rapidly as it is continuously produced and consumed in the TCA cycle. There is likely to be a much greater portion of the total malate in the vacuole where turnover is very slow and there will also be malate in the cytosol and perhaps other organelles with again, different rates of turnover. This is what is meant by metabolic pools; the concept that, due to enzyme compartmentation and limited permeability of organelle membranes there is not complete equilibration of metabolic intermediates between all compartments. That portion of the whole present in each compartment represents a separate pool.

It should be noted that pool is frequently (and erroneously) used to express the total cellular content of a compound. I will use it to refer to that portion of the total content that is more or less segregated in a compartment and distinguished from the rest by a different rate of turnover. It will be realized that when the cell is rapidly killed and extracted in the usual way for analysis (e.g. in boiling 80% ethanol) all of the pools pre-existing in the cell are combined. Analysis of isolated organelle fractions usually fails to give an accurate measure of the pools of metabolites in these organelles *in vivo* because metabolism continues during isolation in the absence of the usual inputs. For example, when purified mitochondria are analysed they are virtually devoid of acids of the TCA cycle. In the special case of vacuoles, in which metabolites are more stable, direct isolation and analysis has shown that they are a major site, for example, of cellular malate (Buser & Matile, 1977).

The demonstration that metabolites are indeed compartmented in plant cells *in vivo* can be demonstrated readily by the use of labelled compounds. This can be illustrated with uniformly labelled ^{14}C-glucose (Grant & Beevers, 1964; Laties, 1964). If the glucose in the cells of a tissue were all in one compartment or if all of the pools contributing to the total glucose in the cell were in complete equilibrium with each other then the continuous introduction of ^{14}C-glucose would label all of the intracellular glucose equally, i.e., it would all come to the same specific radioactivity (dpm per micromole C). On the other hand if, say, 90% of the glucose were present in the vacuole and 10% in the cytosol, the entering ^{14}C might label only the cytosolic pool. Regardless of the number of steps on the pathway or pool sizes of the intermediates, the specific

Table 1. *Compartmentation of glucose in potato slices (Laties, 1964)*

	Specific radioactivity *cpm per* μmol *C*
Ambient glucose	15 000 000
Extracted glucose	500
Respired CO_2	7500
Fraction of glucose in tissue contributing to respiration	1/15

Aged potato slices were incubated in U-^{14}C-glucose (20 μM). After 10 min glucose was extracted from the tissue and its specific radioactivity determined. The specific radioactivity of the CO_2 respired during the experimental period was determined.

radioactivity (s.r.) of the $^{14}CO_2$ would never be higher than that of the ^{14}C-glucose in the cytosol from which it came. In actual experiments, when ^{14}C-glucose is continuously supplied, the $^{14}CO_2$ reaches a constant specific radioactivity within less than 30 min, at a time when all of the intermediates participating in the pathway, including those in the cytosol and mitochondria, have come to the same s.r. as the cytoplasmic glucose. Now if the cells were extracted after 30 min it would be found that the s.r. of the $^{14}CO_2$ exceeded that of the extracted glucose by a factor of 10, since 90% of the glucose in the cells (that in the vacuole) was unlabelled. In fact we can arrive at a maximum value for the fraction of total glucose present in the turnover pool from the relationship observed between the s.r. of the $^{14}CO_2$ and that of the total extracted glucose (Table 1). Knowing that in such a labelling experiment the s.r. of a product can never exceed that of a precursor we can generalize this result (Fig. 1). From the steady state s.r. of the $^{14}CO_2$ we have a maximum value for the s.r. of each intermediate actually undergoing turnover in the pathway. If the s.r. of any precursor extracted at this time is less than that of the $^{14}CO_2$ we can deduce at once that some of that intermediate is not in equilibrium with that of the turnover pool, i.e. it is physically separate from it in the cell. And the factor by which the measured s.r. of the $^{14}CO_2$ exceeds that of a precursor, e.g. that of an acid in the TCA cycle, gives the proportion of the total not in turnover, i.e. not in the mitochondria. Figure 1 illustrates a general case. Note that the relative sizes of the circles indicate the total amount of each intermediate in the compartment, and that these values are different for each intermediate. Nevertheless, at equilibrium the s.r. of each of the intermediates will be the same.

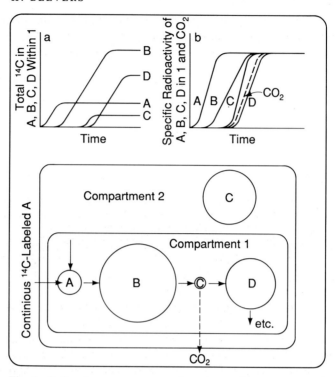

Fig. 1. *Estimating compartmentation.* A reaction sequence A, B, C, D, etc. (e.g. the tricarboxylic acid cycle) is operating with a continuous endogenous supply of A and confined to one compartment (1) in the cell. When U-^{14}C-A in trace amounts is continuously introduced, A, B, C, D, etc. become labelled in sequence (Graph *a*, above). *When equilibrium has been reached* the total amount of ^{14}C in each intermediate will reflect the total mass of each present in the compartment (indicated by the different sizes in the circles in 1 and the ordinate in Graph a). At this time, the specific radioactivity (cpm per µmole C) of each precursor within the compartment and any product such as CO_2 released from the compartment will be the same (Graph *b*, above). So long as ^{14}C-A continues to enter the sequence the specific radioactivity of no intermediate or product will be greater than that of A in the compartment, and the specific radioactivity of any precursor can be determined from knowing that of any later intermediate or product such as CO_2. However if, as indicated, there is a fraction of the total content of intermediate C present elsewhere in the cell, for example, as shown in compartment 2 it will mix with the labelled C from 1 on extraction and the *observed* specific radioactivity will then be lower than the product CO_2. The proportion by which the specific radioactivity of C is lower than that of the product gives the proportion of C that is outside compartment 1 in the cell.

Table 2. *Percentages of individual acid estimated to be in turnover pools in the tricarboxylic acid cycle (MacLennan, Beevers & Harley, 1963)*

Acid	Maize Root tip (2–3 mm)	Maize Root third cm	Maize Coleoptile	Carrot Root discs	Wheat Leaf	Bryophyllum Leaf
Citric	44	13	30	50	15	12
Aconitic	32	9	16	–	2	–
Succinic	89	44	34	–	75	–
Malic	62	36	39	5	23	29

Tissues were incubated in 1-^{14}C-acetate, the respired $^{14}CO_2$ was monitored and samples were killed and extracted at intervals. The organic acids were separated and the s.r. of their carboxyl groups determined. The percentage of acid in turnover was calculated from the relationship between the s.r. of the acids to the (corrected) s.r. of the $^{14}CO_2$ being respired at 2 h.

Experiments of this kind with many different tissues have given clear proof that there exist in plant cells separate pools of metabolites such as those of sugars, amino acids and organic acids. It is usual to assign a large non-turnover pool to the vacuole since this is the largest organelle in most cells, but it is also clear that within the cytoplasm itself separate pools can be recognized by labelling techniques. From a series of experiments in which 1-^{14}C-acetate was provided to plant materials, the fraction of each of the TCA cycle acids undergoing turnover, i.e. in the mitochondria, was calculated by comparing the s.r. of $^{14}CO_2$ and the carboxyl groups of extracted acids (MacLennan, Beevers & Harley, 1963). Some of these results are summarized in Table 2 from which it is clear that for particular acids by far the greatest amount is not undergoing turnover. In different tissues there are large pools of different acids not directly accessible to the corresponding acid labelled in the TCA cycle. The greatest fraction of the total undergoing turnover is seen in the youngest cells of the meristematic region, where vacuoles have not developed and this fraction decreases in progressively older root segments as the vacuolar volume increases dramatically.

Three different pools of malate were demonstrated in corn roots by a double labelling pulse technique (Lips & Beevers, 1966). ^{14}C-labelled HCO_3^- and tritium-labelled acetate were supplied for a period of 15 min and then withdrawn (Fig. 2). Total malate and its ^{3}H and ^{14}C content were monitored for 3 h after the pulse. It was found that ^{3}H-labelled malate (that generated in the TCA cycle in the mitochondria) turned over quite rapidly (t$\frac{1}{2}$~40 min) while that of ^{14}C-malate labelled through the

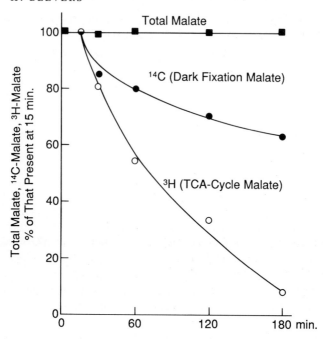

Fig. 2. *Demonstration of different pools of malate in corn roots* (Lips & Beevers, 1964). Corn root segments were exposed to a 15 min pulse of ^{3}H-acetate (to label acids in the tricarboxylic acid (TCA) cycle) plus H^{14}CO$_3^-$ (to label malate produced by CO$_2$ fixation) and transferred to unlabelled medium (0 time). At intervals samples were removed and the total amount of malate in the roots and its ^{14}C and ^{3}H content determined.

CO$_2$ fixation in the cytosol turned over much more slowly ($t\frac{1}{2}>180$ min) while throughout the experimental period the total malate content, presumably largely vacuolar, remained unchanged. The cytosolic malate, but not the mitochondrial, was readily labelled by ^{14}C-malate added from outside.

Thus, the existence of separate pools of various metabolites *in vivo* has been deduced from the kinetic behaviour of labelled intermediates (for a review see Oaks & Bidwell, 1970). Other independent ways of doing this include

(a) ^{31}P-NMR techniques which separately probe the phosphate content and pH of cytosol and vacuole (see Douce *et al.*, this volume), and ^{13}C-NMR analysis of e.g. the malate pools (Roberts, 1984).

(b) Efflux experiments in which ions and metabolites originally present in cytoplasm and vacuole are recognized from kinetic analysis (MacRobbie, 1971, Osmond & Laties, 1969).

(c) Experiments such as that of Delmer (1979) in which the addition of a suitable concentration of dimethylsulfoxide led to the selective release of cytoplasmic tryptophan and reducing sugars, while leaving the vacuolar pools essentially undisturbed.

Isolation of organelles

Although there are other methods by which the intracellular localization of individual enzymes can be investigated, the method that has been most widely used is the isolation of a purified organelle fraction and the demonstration that this organelle contains the bulk of the enzyme activity of interest. This requires that the organelle in the fraction be identified by electron microscopy with the corresponding structure seen in sections of cells and should be supported by a balance sheet showing total enzyme activity in the original extract and in all subsequent fractions. Moreover the purity of the organelle fraction should be established by showing that enzymes typical of other organelles are not present in significant amounts.

Over the past 30 years a very large amount of information has been collected about organelle separation from a variety of plant tissues. Fortunately there are now available several good compendia of detailed information about the separation and properties of plant organelles (Tolbert, 1980; Hall & Moore, 1983; Linskens & Jackson, 1985). Price (1974) produced a valuable methodological survey as well as a book on centrifugation procedures (Price, 1982). Two important publications appeared in 1979; a report edited by Reid of a conference on centrifugal and other methods of organelle separation which also includes recommendations on presentation of data, and a particularly thorough review by Quail (1979) which with 350 references remains the best overall assessment of plant cell fractionation.

No attempt is made here to review this enormous field or to discuss in detail the properties of individual organelles. There is no universal method for organelle isolation that will be successful with all tissues; each presents its own problems. In what follows I outline some general principles that have emerged and illustrate these briefly with examples from our own experience with a particularly propitious tissue, the endosperm of young castor bean seedlings. This tissue has yielded a greater variety of purified organelle fractions than any other.

From the information accumulated on purified organelles it has become clear that particular enzymes are always associated with particular organelles – the concept of marker enzymes (Quail, 1979; Nagahashi, 1985). The best markers for *intact* organelles within a gradient are those enzymes from the matrix; enzymes present only in membranes will be present in ghosts or membrane vesicles generated by breakage of the organelle as well as in intact organelles. Thus, fumarase is frequently used as a marker for intact mitochondria in preference to cytochrome oxidase. Marker enzymes for other organelles or structures in plant cells are listed by Quail (1979).

Cell breakage and grinding media

For organelle isolation, the first requirement is to break open the cells of the tissue of choice in such a way that the organelles suffer minimal damage. It is indeed ironic that the features that most distinguish plant cells from mammalian ones, namely the rigid cell wall and large vacuole, are the very ones that make organelle isolation most difficult. The chopping or grinding force that is necessary to break the cell wall usually leads inevitably to damage, particularly to the more fragile organelles. In practice a compromise is reached; the gentlest grinding procedure that gives a workable yield of extract with the minimum amount of breakage of organelles is adopted. To minimize degradative reactions that supervene once the cells of the tissue have been broken, it is essential to work in the cold and to achieve separation of organelles in the shortest possible time. Various chopping and grinding devices have been used for tissue disintegration; we prefer chopping with new razor blades, which opens cells with the least amount of shear (Beevers, 1975). A wide range of media has been used; the following are the commonest constituents:

(a) an osmoticum, usually sucrose or mannitol ~0·5 M to prevent bursting of organelles due to water uptake through their differentially permeable membranes;

(b) a buffer to maintain pH~7·0. Sufficient buffer capacity must be employed to counteract the frequently very acid contents of the vacuoles.

(c) EDTA to bind heavy metals;

(d) a reducing agent such as dithiothreitol to prevent the oxidation of SH-groups and minimize the generation of oxidants when enzymes and substrates from different compartments are brought into contact;

(e) where phenols and their oxidation products are a particular problem, the addition of polyvinyl pyrolidone (or in solid form Polyclar or PVP) is frequently of benefit (Loomis, 1974);

(f) inhibitors of endogenous proteases and other hydrolytic enzymes, and BSA, particularly in its fat free form which binds free fatty acids, are sometimes used.

The choice of grinding medium and procedure is determined for each tissue by trial and error. Some of the purest organelle fractions have been obtained by starting with large amounts of tissue and using a grinding procedure that breaks only a small fraction of the cells (Douce, 1985). The extent of cell breakage during grinding can be estimated by comparing the activity of an enzyme in the crude extract with the total obtained by exhaustive extraction from the same weight of tissue.

Differential centrifugation

The brei obtained on grinding is usually passed through gauze or nylon net to screen out unbroken tissue and then centrifuged at $\sim 200 \times g$. This step is included in order to remove unbroken cells, cell wall fragments and other debris, but it should be noted that large structures such as starch grains, protein bodies, nuclei and even some plastids also may be sedimented.

Organelle separation from the resulting supernatant or crude extract usually exploits the fact that the individual classes of organelles behave differently in a centrifugal field due to differences in size and density. It should be noted that in most standard extraction procedures some structures are inevitably disrupted. Thus the cell wall, plasmalemma, the large central vacuole and its tonoplast, the ramifying network of endoplasmic reticulum, the cytoskeleton and, frequently, the nucleus are fragmented, even when the bulk of organelles such as mitochondria or microbodies remain intact. However, the broken pieces of membrane from plasmalemma, tonoplast and endoplasmic reticulum apparently reseal immediately into vesicles. Breakage of other organelles results in release of matrix enzymes and the generation of additional membranous material.

Separation of the crude extract into major fractions is achieved by differential centrifugation as introduced initially for liver preparations (see Price, 1982). On centrifuging at *ca.* $10\,000 \times g$ for 30 min a pellet is obtained that is enriched in the larger organelles. Mitochondria are a major component of such pellets and they are frequently referred to and used as 'mitochondrial pellets'. However, in addition to mitochondria, the $10\,000 \times g$ pellet also contains plastids, microbodies and some Golgi vesicles.

When the supernatant solution from this procedure is centrifuged at $100\,000 \times g$ for 1–2 h the resulting pellet is comprised of ribosomes and other small particles as well as membranous vesicles derived from many sources (see above), including the Golgi and endoplasmic reticulum (Tata, 1972). This pellet is referred to as the microsomal pellet, and its heterogeneous nature is immediately revealed by electron microscopy. The supernatant solution remaining after centrifuging at $100\,000 \times g$ contains soluble constituents from the cytosol as well as those released from the breakage of organelles. A balance sheet which follows the distribution of enzyme activity from the crude extract through each stage of centrifugation is essential to establish whether a particular enzyme is associated with a pellet, what fraction of the total is solubilized, and what losses of total enzyme occur during the overall procedure (Reid, 1979).

Some partial purification of the components in the $10\,000 \times g$ and $100\,000 \times g$ pellets may be achieved by resuspension and/or additional steps of differential centrifugation although it should be noted that even the gentlest resuspension and the pelleting itself may lead to organelle breakage.

Separation on sucrose density gradients

By far the most used procedure for further resolution of organelles exploits the fact that the different organelles have different intrinsic densities, since they have distinctive proportions of protein and lipid in their structures (Price, 1982). Centrifugation down a sucrose gradient of increasing density was first employed successfully by Brakke (1951) for the isolation of a plant virus but reached its flowering for organelle isolation in the hands of de Duve and his colleagues (de Duve, 1987).

When a mixture of organelles such as that present in a resuspended $10\,000 \times g$ pellet is placed on top of a 30–60% sucrose gradient and centrifuged at high speed, the individual organelles sediment through the gradient until they arrive at a point equal to their own specific density, where they accumulate. Thus assays for protein and marker enzymes for intact mitochondria show a sharp peak at density $\sim 1 \cdot 18\,\text{g/cm}^3$ and examination of this band in the electron microscope reveals mitochondria with greatly reduced contamination by other organelles. The other major components, Golgi, plastids and microbodies accumulate at other regions of the gradient; if equilibrium has been reached (2–4 h) the gradient is said to be isopycnic. The equilibrium density achieved by each class of organelles on such a gradient is also influenced by the degree to which sucrose penetrates the organelle. Intrinsic specific densities are reached in gradients constructed of non-penetrating materials, such as Percoll, which also avoid some of the osmotic effects of substances of lower molecular weight such as sucrose (Price, 1982). It should also be noted that if a density gradient is constructed *over* a suspension of mixed organelles the organelles will float into the gradient and assume equilibrium positions after suitably prolonged centrifugation.

An additional property that can be exploited in gradient separations is the *rate* at which organelles move through the gradient to their equilibrium position. Thus, of the components present in a $10\,000 \times g$ pellet the fastest moving (highest sedimentation velocity) are the plastids. During a short period of centrifugation, well before the various organelles have reached their equilibrium position in the gradient, the plastids are the major component of the descending front of organelles and a partial separation can be achieved (Miflin & Beevers, 1974). In other preparations, a two-step procedure, exploiting characteristics of sedimentation velocity and equilibrium density has been used for the separation of Golgi (Ray, Shininger & Ray, 1969).

Enzyme assays across a gradient to which a resuspended pellet has been applied usually show that marker enzymes are present at the top of the gradient as well as in intact organelles within the gradient. The relative amount of enzyme not entering the gradient reflects breakage of organelles that occurred during and subsequently to resuspension of the pellet. Some of the damage that results from pelleting at high speed can be avoided by centrifuging onto a cushion of sucrose of appropriate density, but both problems can be greatly alleviated if the crude homogenate rather than a derived pellet is applied directly to the sucrose gradient. On such a gradient the components of the $10\,000 \times g$ pellet as well as those of the microsomal pellet in the previous procedure are simultaneously separated and cytosolic enzymes remain at the top of the gradient. Figure 3 shows such a preparation from castor bean endosperm (Lord, Kagawa & Beevers, 1972). Assays for marker enzymes across the gradient and electron microscopy revealed the identity of the bands as indicated and the amounts of matrix enzymes present at the top of the gradient showed that very little breakage of mitochondria and glyoxysomes had occurred. Components that previously would have been present in the microsomal pellet are present higher in the gradient and assays for marker enzymes show that several constituents are resolved. Vesicles derived from the ER membranes are present at density $1 \cdot 12$ g/cm^3 and detached ribosomes at density $1 \cdot 13$ g/cm^3 (not at equilibrium density after the 3 h centrifugation). Higher in the gradient is a band of mixed membranous material including some membranes from the lipid bodies or spherosomes and at the surface are the spherosomes themselves.

Inclusion of Mg^{++} in the gradient and grinding medium has little effect on the distribution of other organelles but drastically alters that of the ER membranes, which are then recovered with ribosomes still attached

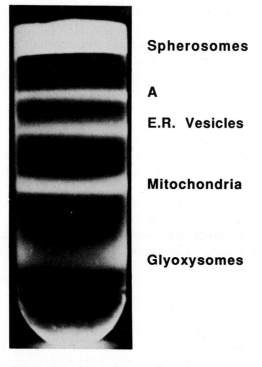

Spherosomes

A

E.R. Vesicles

Mitochondria

Glyoxysomes

Fig. 3. *Separation of organelles of castor bean endosperm tissue on a sucrose gradient* (Lord, Kagawa & Beevers, 1972). A crude extract was placed on a sucrose gradient and centrifuged at high speed for 2 h. The organelles were identified by markers and by electron microscopy. Band A is a mixed membrane fraction. In this gradient a band of ribosomes (not visible) was present just below the band of vesicles derived from the endoplasmic reticulum (ER).

(rough ER) as a broader band at mean density $1·18$ g/cm^3 (Lord *et al.*, 1973). This is the classical Mg^{++} shift that has been observed repeatedly in other systems (Jones, 1985).

Microsomal pellets also include vesicles derived from plasmalemma and tonoplast. Density gradients of sucrose have been used extensively for partial separation of such sealed vesicles in which proton transport coupled to ATP hydrolysis occurs. Different responses of the ATPases to added ions and inhibitors are used as markers (Sze, 1985).

For the isolation of some of the other components inevitably damaged during conventional grinding, isolated protoplasts have been of great value. These are usually prepared from the parent tissue by digestion of the cell walls with mixtures of fungal enzymes in a suitable osmoticum to prevent bursting. The protoplasts are usually collected by centrifugation and then they can be gently broken by osmotic shock or passage through an orifice. Under good conditions intact vacuoles survive and can be separated from the crude extract by centrifugal procedures (Marty, Branton & Leigh, 1980). Intact plastids can also be recovered from such extracts whereas their recovery in other procedures is low (Nishimura & Beevers, 1978a).

In the foregoing, emphasis has been placed on the use of density gradients, particularly linear gradients, for organelle separation. Frequently workers introduce steps in their gradients. When these steps are introduced at densities known from linear gradients to be appropriate, they can be employed usefully for sharpening an organelle band. But their indiscriminate use is to be avoided, for whenever a step is introduced, some material, not necessarily pure, is likely to accumulate at the breakpoint; a rule of thumb is that one can artificially generate any number of bands in a gradient by introducing that number of steps.

Enzyme localization by microscopy

Procedures for the identification of cellular components by microscopic techniques of cytochemistry have been used for many years and the intracellular localization of some enzymes, e.g. phosphatases, by light microscopy was possible if an opaque reaction product could be immobilized (Hall, 1978). Correspondingly, electron cytochemistry based on the generation of electron dense products has provided valuable information. Perhaps the best example of this is the identification of microbodies of various types using the DAB (diamino benzoate) procedure for catalase (Newcomb, 1980).

More recently it has been possible to identify enzyme molecules themselves in sections of suitably fixed material by immunoelectron microscopy. An antibody raised against a pure enzyme (or other antigen of choice) from the plant material is applied to sections prepared for electron microscopy and an antigen–antibody complex is formed. If the antibody is previously conjugated with ferritin the complex is electron dense. Alternatively, the antigen–antibody complex can be recognized by reaction with gold particles coated with protein A. Such techniques, with appropriate controls and objective sampling of electron micrographs can provide important information about the association of proteins with particular organelles *in vivo* (e.g. Titus & Becker, 1985).

Conclusions

Figure 4 summarizes in skeleton form what has been established, from the enzymatic composition of organelle fractions, about metabolic compartmentation in one tissue, the endosperm of young castor bean seedlings (Beevers, 1975, 1982). This is a highly specialized tissue. From the dry, ungerminated condition the life span of the endosperm is about seven days, no cell division occurs and the metabolism is dominated by the massive conversion of stored fat to sucrose. Nevertheless, the results can be used to illustrate some important principles which apply to plant cells more generally.

(1) Degradation and synthesis of polymers occur simultaneously in the same cell and are segregated in different compartments. As shown, the β-oxidation of fatty acids occurs in the glyoxysomes while synthesis of specific fatty acids occurs in the plastids. β-oxidation is now recognized as a general property of microbodies of all higher plant cells (Gerhardt, 1986), and the plastids are the principal or only site of *de novo* fatty acid synthesis from acetyl CoA (Zilkey & Canvin, 1972; Vick & Beevers, 1978). Protein synthesis takes place on the ribosomes in the cytosol (and elsewhere in the cell) while breakdown of storage proteins to amino acids is confined to the vacuole.

(2) Several organelles participate in the major metabolic event and transport of specific metabolites between compartments must occur. In Fig. 4 enzyme sequences in spherosomes, glyoxysomes, mitochondria and cytosol participate in succession in the conversion of stored fat to sucrose. This information defines which metabolites must move from one compartment to another, e.g. succinate from glyoxysomes to mitochondria. Transfer of reducing equivalents by shuttle systems (Mettler & Beevers, 1980) and adenylate transport also occur. In other kinds of cells other organelles play central roles, e.g. the Golgi in cells engaged in cell wall synthesis or secretion (Mollenhauer & Morré, 1980), and the plastids in amino acid biosynthesis (Miflin, 1974; Miflin & Lea, 1980), and the storage and breakdown of starch (Preiss, 1988). The elucidation of the transport steps in all such systems poses a major problem for the future. As indicated earlier, investigations on mitochondrial and chloroplast transporters are leading the way (Heldt & Flügge, 1987).

(3) Organelles not participating in major metabolic pathways have essential roles in the synthesis and maintenance of enzymes and other organelles. Thus in Fig. 4 the hydrolysis of protein in the vacuole provides the amino acids for the ribosomal synthesis of enzymes and other proteins used in the biogenesis of glyoxysomes and mitochondria, whose numbers increase dramatically in the first days of growth. Fatty acids are produced

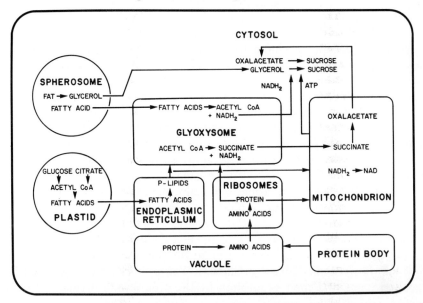

Fig. 4. *Major functions of organelles in the endosperm tissue of young castor bean seedlings.* All of the organelles shown have been isolated in a highly purified and functional state. We have no experimental evidence for the existence of a nucleus in these cells but remain confident that one is present.

in the plastids and are used for the synthesis of all the phospholipids in the endoplasmic reticulum (Beevers, 1975; Moore, 1982). The Golgi is also a major site of phospholipid synthesis in some cells (Montague & Ray, 1977). The full elucidation of organelle biogenesis remains a problem for the future, although important beginnings have been made for chloroplasts and mitochondria. These organelles have their own genomes and machinery for protein synthesis, which is responsible for a small fraction of their total protein complement; they show cytoplasmic inheritance. Their progenitors are passed on to the zygote through the gamete(s). Thus, for these organelles there is always a recognized target for the acquisition of newly synthesized components as they grow and divide. Synthesis *de novo* of organelles (in the absence of progenitor organelles) poses additional problems.

For all membrane-bound organelles biogenesis requires (a) the synthesis of a distinctive set of phospholipids for their membranes, (b) the synthesis of membrane and matrix proteins, and (c) an assembly mechanism that ensures a perfect match, such that the proteins of a particular organelle (and no other proteins) are found there and in the correct topographical arrangement.

The endoplasmic reticulum has an important role in organelle biogenesis. The membranes of some organelles, e.g. Golgi and vacuoles are apparently derived directly from membranes of the endoplasmic reticulum, and attached ribosomes are responsible for the synthesis and insertion of some storage and secretory proteins (Chrispeels, 1980). However they are assembled, all organelles appear to depend on the endoplasmic reticulum for the production of at least some of their phospholipids (Moore, 1982). Carrier proteins may be responsible for their transport (Kader, 1985) although it is not clear how specificity is imposed. It is clear that the endoplasmic reticulum must take part in the core glycosylation of any glycoprotein in other organelles (Beevers & Mense, 1977; Chrispeels, 1980; Jones, 1985) and the Golgi is the site of further elaboration of such molecules (Mollenhauer & Morré, 1980). Synthesis on cytoplasmic ribosomes appears to be the norm, both for membrane and matrix proteins of most organelles. The processing and sorting of these proteins and their acquisition by the appropriate organelles is a highly specific and selective process which is now under active investigation.

Acknowledgement

Over the years the work in my group has been supported by grants from the National Science Foundation, most recently by grant PCM 84-03542.

References

Akazawa, T. & Hara-Nishimura, I. (1985). Topographic aspects of biosynthesis, extracellular secretion, and intracellular storage of proteins in plant cells. *Ann. Rev. Plant Physiol.* **36**, 441–72.

Beevers, H. (1975). Organelles from castor bean seedlings: biochemical roles in gluconeogenesis and phospholipid metabolism. In *Recent Adv. Chem. Biochem. of Plant Lipids*, ed. T. Galliard & E. I. Mercer, pp. 287–99. London: Academic Press.

Beevers, H. (1982). Fat metabolism in seeds: role of organelles. In *Biochemistry and Metabolism of Plant Lipids*, ed. J. G. M. Wintermans & P. J. L. Kuiper, pp. 223–35. Amsterdam: Elsevier.

Beevers, L. & Mense, R. (1977). Glycoprotein synthesis in cotyledons of *Pisum sativum* L. *Plant Physiol.* **60**, 703–8.

Brakke, M. K. (1951). Density gradient centrifugation. A new separation technique. *J. Am. Chem. Soc.* **73**, 1847–8.

Buser, C. & Matile, P. (1977). Malic acid in vacuoles isolated from *Bryophyllum* leaf cells. *Z. Pflphysiol.* **82**, 462–6.

Chrispeels, M. J. (1980). The endoplasmic reticulum. In *The Biochemistry of Plants*, vol. 1, ed. P. K. Stumpf & E. E. Conn, pp. 389–412. New York: Academic Press.

Delmer, D. P. (1979). Dimethylsulfoxide as a potential tool for analysis of compartmentation in living plant cells. *Plant Physiol.* **64**, 623–9.

de Duve, C. (1987). Exploring cells with a centrifuge. *Science* **189**, 186–94.

Douce, R. (1985). *Mitochondria in Higher Plants*. New York: Academic Press.

Gerhardt, B. (1986). Basic metabolic functions of the higher plant peroxisome. *Physiol. Vég.* **24**, 397–410.

Grant, B. R. & Beevers, H. (1964). Absorption of sugars by plant tissues. *Plant Physiol.* **39**, 78–85.

Hall, H. L. (1978). *Electron Microscopy and Cytochemistry of Plant Cells*. New York: Elsevier/North Holland.

Hall, J. L. & Moore, A. L. (1983). *Isolation of Membranes and Organelles from Plant Cells*. New York: Academic Press.

Harley, S. M. & Beevers, H. (1982). Ricin inhibition of *in vitro* protein synthesis by plant ribosomes. *Proc. Natn. Acad. Sci. USA* **79**, 5935–8.

Heldt, H. W. & Flügge, U. I. (1987). Subcellular transport of metabolites in plant cells. In *The Biochemistry of Plants*, vol. 12, ed. P. K. Stumpf & E. E. Conn, pp. 50–86. New York: Academic Press.

Jones, R. L. (1985). Endoplasmic reticulum. In *Cell Components*, ed. H. F. Linskens & J. F. Jackson, pp. 304–30. Berlin: Springer.

Kader, J. C. (1985). Lipid-binding proteins in plants. *Chem. Phys. Lipids* **38**, 51–8.

Laties, G. G. (1964). The relation of glucose absorption to respiration in potato slices. *Plant Physiol.* **39**, 391–7.

Linskens, H. F. & Jackson, J. F. (1985). *Cell Components*, vol. 1, Modern Methods of Plant Analysis, New Series. Berlin: Springer.

Lips, S. H. & Beevers, H. (1966). Compartmentation of organic acids in corn roots. II. Differential labeling of two malate pools. *Plant Physiol.* **44**, 713–17.

Loomis, W. D. (1974). Overcoming problems of phenolics and quinones in the isolation of plant enzymes and organelles. *Meth. Enzymol.* **31**, 528–44.

Lord, J. M., Kagawa, T. & Beevers, H. (1972). Intracellular distribution of enzymes of the cytidine disphosphate choline pathway in castor bean endosperms. *Proc. Natn. Acad. Sci. USA* **69**, 2429–32.

Lord, J. M., Kagawa, T., Moore, T. S. & Beevers, H. (1973). Endoplasmic reticulum as the site of lecithin formation in castor bean endosperm. *J. Cell Biol.* **57**, 659–67.

MacLennan, D. H., Beevers, H. & Harley, J. (1963). Compartmentation of acids in plant tissues. *Biochem. J.* **89**, 316–27.

MacRobbie, E. A. C. (1971) Fluxes and compartmentation in plant cells. *Ann. Rev. Plant Physiol.* **22**, 75–96.

Marty, F., Branton, D. & Leigh, R. A. (1980). Plant vacuoles. In *The Biochemistry of Plants*, vol. 1, ed. P. K. Stumpf & E. E. Conn, pp. 625–58. New York: Academic Press.

Mettler, I. J. & Beevers, H. (1980). Oxidation of NADH in glyoxysomes by a malate-aspartate shuttle. *Plant Physiol.* **66**, 555–60.

Miflin, B. J. (1974). The location of nitrite reductase and other enzymes related to amino acid biosynthesis in the plastids of roots and leaves. *Plant Physiol.* **54**, 550–5.

Miflin, B. J. & Beevers, H. (1974). Isolation of intact plastids from a range of plant tissues. *Plant Physiol.* **53**, 870–4.

Miflin, B. J. & Lea, P. J. (1980). Ammonia assimilation. In *The Biochemistry of Plants*, vol. 5, ed. P. K. Stumpf & E. E. Conn, pp. 169–202. New York: Academic Press.

Mollenhauer, H. H. & Morré, D. J. (1980). The Golgi apparatus. In *The Biochemistry of Plants*, vol. 1, ed. P. K. Stumpf & E. E. Conn, pp. 437–88. New York: Academic Press.

Montague, M. I. & Ray, P. M. (1977). Phospholipid-synthesizing enzymes associated with Golgi dictyosomes from pea tissue. *Plant Physiol.* **59**, 225–30.

Moore, T. S. (1982). Phospholipid synthesis. *Ann. Rev. Plant Physiol.* **33**, 235–59.

Nagahashi, G. (1985). The marker concept in cell fractionation. In *Cell Components*, ed. H. F. Linskens & J. F. Jackson, pp. 66–84. Berlin: Springer.

Newcomb, E. H. (1980). The general cell. In *The Biochemistry of Plants*, vol. 1, ed. P. K. Stumpf & E. E. Conn, pp. 1–54. New York: Academic Press.

Nishimura, M. & Beevers, H. (1978a). Isolation of intact plastids from protoplasts from castor bean endosperm. *Plant Physiol.* **62**, 40–3.

Nishimura, M. & Beevers, H. (1987b). Hydrolases in vacuoles from castor bean endosperm. *Plant Physiol.* **62**, 44–8.

Oaks, A. & Bidwell, R. G. S. (1970). Compartmentation of intermediary metabolites. *Ann. Rev. Plant Physiol.* **21**, 43–66.

Osmond, C. B. & Laties, G. G. (1969). Compartmentation of malate in relation to ion absorption in beet. *Plant Physiol.* **44**, 7–14.

Preiss, J. (1988). Biosynthesis of starch and its regulation. In *The Biochemistry of Plants*, vol. 14, ed. P. K. Stumpf & E. E. Conn, pp. 181–254. New York: Academic Press.

Price, C. A. (1974). Plant cell fractionation. *Meth. Enzymol.* **31**, 501–19.

Price, C. A. (1982). *Centrifugation in Density Gradients*. New York: Academic Press.

Quail, R. H. (1979). Plant cell fractionation. *Ann. Rev. Plant Physiol.* **30**, 425–84.

Raven, J. A. (1987). The role of vacuoles. *New Phytol.* **106**, 357–422.

Ray, P. M., Shininger, T. L. & Ray, M. M. (1969). Isolation of β-glucan synthetase particles from plant cells and identification with Golgi membranes. *Proc. Natn. Acad. Sci. USA* **64**, 605–12.

Reid, E. (1979). *Plant Organelles.* Chichester: Ellis Harwood.

Roberts, J. K. M. (1984). Study of plant metabolism *in vivo* using NMR-spectroscopy. *Ann. Rev. Plant Physiol.* **35**, 375–86.

Saunders, J. A. & Conn, E. E. (1978). Presence of the cyanogenic glucoside Dhurrin in isolated vacuoles from sorghum. *Plant Physiol.* **61**, 154–7.

Srere, P. (1987). Complexes of sequential metabolic enzymes. *Ann. Rev. Biochem.* **56**, 89–124.

Sze, H. (1985). H^+-translocating ATPases: advances using membrane vesicles. *Ann. Rev. Plant Physiol.* **36**, 175–208.

Tata, J. R. (1972). Preparation and properties of microsomal and sub-microsomal fractions from secretory and non-secretory tissues. In *Subcellular Components*, ed. G. D. Birnie, pp. 185–213. London: Butterworth.

Titus, D. E. & Becker, W. M. (1985). Investigation of the glyoxysome-peroxisome transition in germinating cucumber cotyledons using double-label immunoelectron microscopy. *J. Cell Biol.* **101**, 1288–99.

Tolbert, N. E. (1980). *The Plant Cell. The Biochemistry of Plants*, vol. 1, ed. P. K. Stumpf & E. E. Conn. New York: Academic Press.

Vick, B. & Beevers, H. (1978). Fatty acid synthesis in endosperm of young castor bean seedlings. *Plant Physiol.* **62**, 173–8.

Zilkey, B. F. & Canvin, D. T. (1972). Localization of oleic acid biosynthesis enzymes in the proplastids of developing castor bean endosperm. *Can. J. Bot.* **50**, 323–6.

J. L. HARWOOD

Fatty acid and lipid biosynthesis

Introduction

It is sometimes forgotten by plant biochemists, in the enthusiasm of their research discoveries on enzymes, that metabolism takes place within complicated compartments of the eukaryotic cell. Thus, the individual reactions involved in a metabolic scheme may depend on all sorts of complex interactions or transport processes which may limit overall rates more than the potential activity of the enzymes themselves. For lipid metabolism, there are many important examples of cooperation of compartments for metabolic sequences. It is only relatively recently that all the subtleties of such interactions have been considered seriously and, even now, there has been amazingly little work in some areas (such as movements of lipids between compartments). In this short review I shall describe the current state of our knowledge and point out a few subjects deserving of urgent attention.

Fatty acid synthesis

Most lipids contain acyl chains and it is these fatty acids which give lipids their characteristic hydrophobic and/or amphipathic properties. Accordingly I shall discuss fatty acid synthesis, and the role of cellular compartmentation in this process, first.

De novo synthesis and the plastid

In leaf tissues it has been suggested that the chloroplast is the only site of fatty acid synthesis *de novo* (Weaire & Kekwick, 1975; Ohlrogge, Kuhn & Stumpf, 1979). One can, perhaps, extrapolate to non-green tissues and cite a pre-eminent role for the plastid. Certainly, I have little doubt that the plastid is the *major* subcellular location for fatty acid synthesis *de novo* but whether it is the *exclusive* site is much less certain. A number of observations (see Harwood, 1988) suggest that some formation of fatty acids *de novo* takes place in other parts of the cell. Indeed, although acyl carrier protein (ACP) could only be detected immunologically in chloroplasts from spinach leaves (Ohlrogge *et al.*, 1979) measurement of ACP using acetyl-CoA:ACP transacylase showed that small amounts were present outside the plastid in tissues such as germinating peas and maturing avocado fruits (Roberto, F., Harwood, J. L. & Stumpf, P. K., unpublished data). Nevertheless, notwithstanding the above caveats one can say with confidence that the plastid is generally the site for *de novo* fatty acid synthesis.

Two major enzyme complexes are used for the formation of the major saturated fatty acid, palmitate. These are a multifunctional protein, acetyl-CoA carboxylase and a dissociable Type II complex, fatty acid synthetase (Harwood, 1988). Recent improvements in isolation procedures (mainly involving the use of effective proteinase inhibitors and a rapid affinity chromatographic step on avidin-sepharose) have resulted in the isolation of acetyl-CoA carboxylases of 220–240 kDa from several plant tissues. For a representative method see Slabas & Hellyer (1985) who purified the enzyme from rape seed. In animal tissues, acetyl-CoA carboxylase is considered, usually, to control the overall rate of fatty acid (and triacylglycerol) synthesis and is subject to a number of regulatory factors (Hardie, Carling & Slim, 1989). The methods of regulating mammalian acetyl-CoA carboxylase (tricarboxylic acid or acyl-CoA levels, phosphorylation/dephosphorylation) do not appear to have a role in controlling the plant enzyme. Only two methods of regulating activity have been demonstrated and only one of these applies to non-photosynthetic tissues. In leaves it is well known that light causes an increase of 20–25 times the dark-grown acetyl-CoA carboxylase activity. Most likely, this elevation in activity is caused by a combination of changes in stromal pH and the concentrations of Mg^{++}, ATP and ADP on illumination. Changes in these parameters commensurate with physiological values were shown to cause a 24-fold increase in the activity of maize acetyl-CoA carboxylase *in vitro* (Nikolau & Hawke, 1984). The other method of control is by changes in acetyl-CoA carboxylase protein levels. A good example is shown for developing oil-seed rape in Fig. 1. Although carboxylase protein levels were not measured specifically in this study, PAGE patterns were in agreement with a large increase in acetyl-CoA carboxylase protein coincident with the rise in activity at the onset of oil deposition (Turnham & Northcote, 1983).

Recently, a number of aryloxyphenoxypropionate and cyclohexanedione herbicides, have been shown to act by inhibiting acetyl-CoA carboxylase (see Harwood, 1988; Walker *et al.*, 1989). The enzyme from

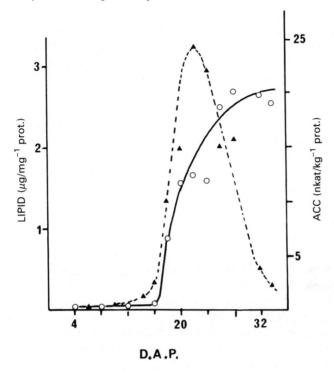

Fig. 1. Developmental profiles for acetyl-CoA carboxylase activity (ACC) (▲) and oil deposition (O) in developing oil-seed rape. D.A.P. = days after pollination. (Data taken from Turnham & Northcote (1983), with permission).

Poaceae but not from other monocotyledons such as Juncaceae or Cyperaceae (Kobek *et al.*, 1988) or from dicotyledons is sensitive. This surprising observation raises interesting questions about the structure of acetyl-CoA carboxylase from different plants. In addition, there is a problem with reconciling experimental results on fatty acid elongation in susceptible plants with the inhibition of acetyl-CoA carboxylase. As will be seen later, elongation of pre-formed fatty acids requires malonyl-CoA co-substrate (see Harwood, 1988). However, when preparations from susceptible plants, treated with aryloxyphenoxypropionate herbicide, are incubated with ^{14}C-acetate the labelling of very long chain fatty acids is hardly affected, although *de novo* synthesis is completely inhibited (Walker, Ridley & Harwood, 1988). This result raises the possibility that in barley, at least, there may be a herbicide-insensitive acetyl-CoA carboxylase outside the plastid.

Fatty acid synthetase contains a number of individual proteins which catalyse the partial reactions as well as ACP. Following pioneering work in three laboratories (Caughey & Kekwick, 1982; Hoj & Mikkelson, 1982; Shimakata & Stumpf, 1983) the various enzymes have been purified and studied (see Stumpf, 1987; Harwood, 1988). Of particular interest has been the question of whether or not any one partial reaction controls the overall rate of synthesis and also how the chain-length of fatty acyl end-products is controlled.

So far as a candidate for overall regulation is concerned then Shimakata & Stumpf (1983) produced evidence to suggest that acetyl-CoA: ACP transacylation was the slowest enzyme reaction. All the other reactions, with the exception of β-ketoacyl-ACP synthetase 1 were considerably faster when measured *in vitro* (Shimakata & Stumpf, 1983). Furthermore, intermediates have never been shown to accumulate – implying that an early reaction is rate-limiting. An interesting recent development has been the discovery of a cerulenin-insensitive aceto acetyl-ACP synthetase (i.e. a very short-chain condensing enzyme). This enzyme can use acetyl-CoA directly as the primer molecule and, since it has about five times the activity of acetyl-CoA:ACP transacylase in spinach leaves, can effectively by-pass the latter reaction (Fig. 2; Jaworski, Clough & Barnam, 1989). In fact, in *E. coli* it has been suggested that acetoacetyl-ACP synthetase can catalyse acetyl-CoA: ACP transacylation as a side-reaction [Jackowski & Rock, 1987]. Thus, the overall rate of fatty acid synthetase could be regulated by either acetoacetyl-ACP synthetase or by β-ketoacyl-ACP synthetase 1 activity *in vivo*.

Although the vast majority of *de novo* synthesis results in the formation of 16- and 18-carbon products, certain economically important crops such as palm kermal and coconut accumulate medium-chain fatty acids. The possible regulatory mechanisms involved have been discussed (Slabas *et al.*, 1984; Pollard & Singh, 1987; Harwood, 1988) but no clear picture has yet emerged. This is an important area for future research.

Once palmitoyl-ACP has been produced by fatty acid synthetase it is chain-lengthened to stearoyl-ACP using a specific cerulenin-insensitive, arsenite-sensitive β-ketoacyl-ACP synthetase 2. So far as we know, it is only the condensing reaction that needs a special enzyme, the other proteins (β-ketoacyl-ACP reductase, β-hydroxyacyl-ACP dehydrase, enoyl-ACP reductase) having a broader chain-length specificity (Harwood, 1988).

Stearate rarely accumulates in plants and the 18-carbon unsaturated

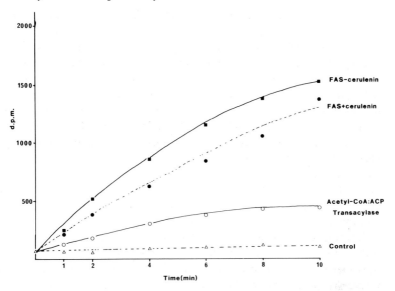

Fig. 2. Comparison of the rate of fatty acid synthesis (FAS) and acetyl CoA:ACP transacylation in oil seed rape homogenates.
Activity measured by the incorporation of radioactivity from [^{14}C]acetyl CoA into acid-insoluble compounds (i.e. acyl-ACPs)
M. Walsh & J. L. Harwood (1989) unpublished data.

acids (oleic, linoleic, α-linolenic) are the major products in plants. The initial desaturation also takes place in the plastid and uses stearoyl-ACP substrate. Stearoyl-ACP Δ9-desaturase has been purified from developing safflower seeds (McKeon & Stumpf, 1982) where it is soluble. In other tissues, the desaturase appears to be loosely associated with membranes (see Weiare & Kekwick, 1975; Walker & Harwood, 1985).

Once oleoyl-ACP has been formed it can have two major fates. First, the oleate may be hydrolysed by stromal oleoyl-ACP thioesterase, (Shine, Mancha & Stumpf, 1976), transferred to coenzyme A by an envelope acyl-CoA synthetase (Joyard & Stumpf, 1981) and exported to the extra-plastid compartments. Second, the oleate may be used as a substrate by glycerol 3-phosphate acyltransferase to provide the initial acyl group at the *sn*-1 position of plastid-synthesized lipids (see Roughan, 1987). These reactions are discussed further in the sections: *The difference between 16:3 and 18:3 plants* and *Possible roles of lipid exchange proteins*.

Modifications to produce unsaturated and very long chain products

The next two desaturations after oleate are at the Δ12- and Δ15-positions. A concensus has emerged over the last decade that the major substrate for Δ12-desaturation is phosphatidylcholine. This was first demonstrated directly by Pugh & Kates (1973) for *Candida lipolytica* and was convincingly shown in higher plants by Stymne & Appelqvist (1978; section on *Acylation of glycerol-3-phosphate*). The diacylglycerol so produced is then incorporated into phosphatidylcholine via cholinephosphotransferase (section *Glycosylglyceride formation*.) Second, an enzyme lysophosphatidylcholine:acetyl-CoA acyltransferase is capable of catalysing the selective incorporation of acids into the *sn*-2 position (Stymne & Stobart, 1987).

Although in some plants at any rate, both positions of phosphatidylcholine are capable of being desaturated (Stymne & Stobart, 1987), in practice because of the concentration of oleate at the *sn*-2 position then Δ-12 desaturation takes place here.

For plants where oleate is retained within the plastid, other lipids must be used for Δ12- (and Δ15) desaturation as phosphatidylcholine is not a regular constituent of the thylakoids (Gounaris, Barber & Harwood, 1986). This would have parallels to the situation in cyanobacteria where all lipids, to various degrees, seem to be substrates for these desaturations (Murata & Nishida, 1987).

For the desaturation of linoleate to α-linolenate (Δ15-desaturation) there are several possibilities (Fig. 3). In most photosynthetic tissues it appears that phosphatidylcholine serves as a donor of linoleate to chloroplast monogalactosyldiacylglycerol which is the actual substrate for Δ15-desaturation (Roughan *et al.*, 1979; Jones & Harwood, 1980). This movement of lipid from the endoplasmic reticulum to the plastid may require lipid-exchange carrier protein (Ohnishi & Yamada, 1982). Alternatively, in plants where desaturation continues within the plastid then monogalactosyldiacylglycerol will probably still be the main substrate (cf. cyanobacterial lipid desaturation rates; Sato & Murata, 1982) but other lipids may also be used. In seed tissues there is a third possibility and that is that phosphatidylcholine is the substrate for Δ15-desaturation. Thus, experiments with developing linseed (one of the few seeds which accumulates much α-linolenate) have implicated phosphatidylcholine as a substrate for both Δ12- and Δ15-desaturation (Stymne, Green & Tonnet, 1989).

Just as the formation of polyunsaturated fatty acids usually involves the cooperation of more than one cell compartment, so the elongation of pre-

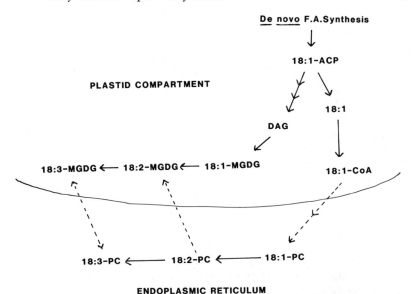

Fig. 3. Possible routes for linolenate synthesis in different plant tissues. −−−=lipid transfer; ACP=Acyl carrier protein; DAG=diacylglycerol; MGDG=monogalactosyldiacylglycerol; PC=phosphatidylcholine

formed acyl chains also takes place outside the plastid. In this case acyl-CoAs (formed by the envelope or cytoplasmic acyl-CoA synthetases) are the usual substrates (see Harwood, 1988). In any given plant tissue there seem to be several elongation systems which are chain-length specific (e.g. Walker & Harwood, 1986; Cassagne et al., 1987). For all the enzyme systems, malonyl-CoA is the source of the 2-carbon addition unit and NADPH or NADH are used as reductants. In the case of leek, attempts have been made to purify elongases, following detergent solubilization (e.g. Agrawal & Stumpf, 1985; Lessire, Bessoule & Cassagne, 1985). Although saturated fatty acids are the usual substrates for elongation, in some seeds very long chain unsaturated acids accumulate. Early studies with *Crambe* (Appleby, Gurr & Nicols, 1974) gave some information on the synthesis of erucate which is a major component (40–60%) of the seed oil fatty acids of early agricultural varieties of rape (Gunstone, Harwood & Padley, 1986). In general, the enzymes involved in unsaturated fatty acid elongation, seem to have similar locations and properties as the more common saturated acid elongases.

Table 1. *The fatty acid composition of thylakoid lipids from representative '16:3' and '18:3'-plants*

Plant	Lipid	Fatty acid composition (% total)					
		16:0	16:1[a]	16 :3	18:1	18:2	18:3
Spinach	MGDG	1	—	25	1	2	72
	DGDG	4	—	5	2	2	87
	SQDG	39	trace	trace	1	7	53
	PG	11	32	—	2	4	47
Barley	MGDG	4	—	—	1	3	90
	DGDG	9	2	—	3	7	78
	SQDG	32	3	—	2	5	55
	PG	18	27	—	2	11	38

[a]Palmitoleate in all lipids except PG where it is mainly
trans-Δ3-hexadecanoate.
Abbreviations: MGDG = monogalactosyldiacylglycerol
    DGDG = digalactosyldiacylglycerol
    SQDG = sulphoquinovosyldiacylglycerol
    PG   = phosphatidylglycerol

The difference between '16:3' and '18:3'-plants

A number of those plants commonly used by biochemists (notably spinach) accumulate *cis*, 7,10,13-hexadecatrienoate (16:3) in their leaf monogalactosyldiacylglycerol. This acid is also found in much smaller amounts in digalactosyldiacylglycerol but is excluded from other leaf lipids (Table 1). The origin and synthesis of this acid are essentially tied to the general method of lipid metabolism in such species. Although experiments to data have been concentrated on leaves, it may be realistic to extrapolate them to non-photosynthetic tissues. Accordingly a brief discussion will be made here and expanded in the section on *Possible roles of lipid exchange proteins*.

For hexadecatrienoic acid to be formed, successive desaturations of palmitate, while esterified to the *sn*-2 position of monogalactosyldiacylglycerol, seem to be involved (Andrews & Heinz, 1987; Sato, Seyama & Murata, 1986; Thompson *et al.*, 1986). Moreover, the palmitate has to be transferred first from palmitoyl-ACP via lysophosphatidate acyltransferase. The latter enzyme has to out-compete the acyl-ACP thioesterase activity of the plastid in order to ensure that the palmitate does not reach the extra-chloroplastic compartment. If it did, then esterification would take place to the *sn*-1 position of glycerol 3-phosphate and the diacylglycerol which eventually resulted would form an inappropriate backbone for desaturation to hexatrienoate in monogalactosyl-diacylglycerol.

One way in which the retention or export of acyl groups could be controlled is by regulating the relative activities of plastid acyl-ACP hydrolysis and glycerol 3-phosphate acylation. Interestingly, the activity of these reactions can be altered according to the nature of the ACP isoform which is esterified (Guerra, Ohlrogge & Frentzen, 1986). Thus, in spinach which has two isoforms, acyl-ACP-1 was used preferentially by oleoyl-ACP thioesterase while acyl-ACP-2 was used by glycerol 3-phosphate acyltransferase. Thus, plants (and genetic manipulators) could possibly change the pattern of lipid metabolism in a tissue by altering the expression of ACP isoforms (see Ohlrogge, 1987; Harwood, 1988).

Acylation of glycerol 3-phosphate

According to the Kornberg-Pricer pathway, the acylation of glycerol 3-phosphate requires two separate enzymes. Indeed, two acyltransferases have been found in plastid and microsomal (probably endoplasmic reticulum) fractions. These enzyme pairs differ from each other not only in location but also in their substrate and positional specificities.

The plastid glycerol 3-phosphate acyltransferase has been purified from several tissues (see Harwood, 1988). In the case of squash, three isoforms have been separated from stromal fractions (Nishida *et al.*, 1987). Although stromal glycerol 3-phosphate acyltransferase will use acyl-CoAs as well as acyl-ACPs, when presented with mixtures a strong preference for acyl-ACP is observed (Frentzen *et al.*, 1983). In addition, as mentioned above, the activity of the enzyme is influenced by the ACP isoform (Guerra *et al.*, 1986). In terms of fatty acid preference, the plastid enzyme shows selectivity for oleoyl-ACP over saturated-ACPs – thus ensuring that oleate is concentrated at the *sn*-1 position (Table 2).

Lysophosphatidate acyltransferase is bound firmly to the plastid envelope membrane. Under normal conditions it appears not to be rate-limiting since lysophosphatidate is rarely detected. Because the substrate is the 1-acyl isomer, it follows that lysophosphatidate acyltransferase can only esterify the *sn*-2 position *in vivo*. Competition experiments have shown that the enzyme uses preferentially palmitate (Joyard, Chuzel & Douce, 1979) and that palmitoyl-ACP is used rather than palmitoyl-CoA (Frentzen *et al.*, 1983). Of course, in practice since the plastid produces palmitoyl-ACP as an end-product of *de novo* fatty acid synthesis then palmitoyl-CoA would be unavailable.

Table 2. *Fatty acid distributions in the phosphatidate generated by plastids or by endoplasmic reticulum*

	Fatty acid content (% total)				
Phosphatidate source	16:0	18:0	18:1	18:2	18:3
Spinach plastid[a]					
sn-1	10·1	—	89·9	—	—
sn-2	96·3	—	3·7	—	—
Safflower[b] endoplasmic reticulum					
sn-1	51	16	20	11	2
sn-2	0	0	4	81	15

[a]=Joyard & Douce (1987) using mixtures of [^{14}C]palmitoyl-ACP and [^{14}C]oleoyl-ACP.

[b]=Griffiths *et al.* (1985) using equimolar mixtures of [^{14}C] acyl-CoAs.

In comparison with experiments examining the plastid enzyme, rather less is known about the acyltransferases responsible for forming phosphatidic acid outside the above organelle. Part of the reason for this deficiency is because both enzymes are present in membranes (of the endoplasmic reticulum) and, therefore, more difficult to study than soluble proteins. Several workers have studied the two acyltransferases in cell-free preparations from various plants and, later, in microsomal fractions (see Stymne & Stobart, 1987). Substrate specificity experiments have sometimes given conflicting results but, when mixtures of acyl-CoA substrates are presented palmitate is preferentially used for glycerol 3-phosphate acylation. Typical results would be those for microsomes from developing safflower seeds. With such preparations, a substrate preference of palmitate>oleate=linoleate>stearate was found (Ichihara, 1984). Furthermore, the lysophosphatidate acyltransferase from safflower shows a strong preference for linoleate over oleate with saturated acyl-CoAs scarcely used at all (Griffiths, Stobart & Stymne, 1985).

Thus, the substrate specificities of the extra-plastid acyltransferases lead to a typical *sn*-1-saturated, *sn*-2-unsaturated distribution of acyl groups in the phosphatidate generated in the endoplasmic reticulum of seed tissues. This distribution contrasts with that in phosphatidate formed from acyl-ACPs in the plastid (Table 2; Harwood, 1989).

Triacylglycerol accumulation

Once phosphatidate has been formed, it can be dephosphorylated to yield diacylglycerol. The enzyme responsible is phosphatidate phosphohydrolase, the properties of which have been reviewed recently (Harwood & Price-Jones, 1987). This enzyme has been studied in several subcellular fractions and it is difficult to know if one can speak of typical locations outside the plastid. In a number of tissues, a major site of phosphatidate phosphohydrolase seems to be the endoplasmic reticulum. Indeed, by analogy with mammalian tissues soluble activity may represent enzyme that has been detached from such membranes. Additionally, phosphatidate phosphohydrolase may be rate-limiting for triacylglycerol accumulation in many situations. It probably does not display much substrate specificity so that the accumulating diacylglycerol has a very similar substrate specificity to the precursor phosphatidate (Harwood, 1989).

The final step in triacylglycerol formation is the diacylglycerol acyltransferase. This enzyme is located in the endoplasmic reticulum and, like the glycerol 3-phosphate acyltransferases, uses acyl-CoA substrates. Diacylglycerol acyltransferase has been studied in a number of plant tissues. In some developing seeds this acyltransferase seems to have a substrate preference which could account for the pattern of acyl moieties at the *sn*-3 position (Cao & Huang, 1986; Harwood, 1989). However, in other tissues such as from peanut (Cao & Huang, 1986) or cocoa (McHenry & Fritz, 1987) the fatty acids at the *sn*-3 position seem to be dependent on the acyl-CoA pool (Harwood, 1989).

A major area of controversy concerning triacylglycerol accumulation lies with the morphology and production of oil bodies. I will not address this subject here but refer the reader to Gurr (1980) and Stymne & Stobart (1987) who have reviewed the various possibilities and arguments.

Glycosylglyceride formation

Glycosylglycerides are formed in plastids but the source of diacylglycerol for this synthesis varies between plants. In '16:3'-plants the diacylglycerol is generated by plastid phosphatidate phosphohydrolase whereas in '18:3'-plants the diacylglycerol is formed by the action of endoplasmic reticulum phosphatidate phosphohydrolase. Because of the differences in the typical acyl composition and distribution for phosphatidate formed in these two locations, the glycosylglyceride backbones are quite characteristic.

Table 3. *Activity of enzymes involved in galactosyldiacylglycerol formation in chloroplasts from pea and spinach*

		Pea	Spinach
Phosphatidate turnover (T1/2, min)		38·5[a]	1·7[a]
Phosphatidate phosphohydrolase (nmol h^{-1} mg^{-1} protein)	(outer env.)	–	7[b]
	(inner env.)	–	60[b]
DAG:UDP-gal. galactosyltransferase (nmol h^{-1} mg^{-1} protein)	(outer env.)	2424[c]	120[b]
	(inner env.)	402[c]	880[b]

[a]Gardiner & Roughan (1983).
[b]Block *et al.* (1983*b*).
[c]Cline and Keegstra (1983).

Taking pea as a typical '18:3'-plant and spinach as a '16:3'-plant, one can easily see the difference in metabolism. The phosphatidate phospho-hydrolase in plastids from spinach has a much higher activity (Table 3) thus ensuring a good supply of diacylglycerol within the organelle (Gardiner & Roughan, 1983). Also, the phosphohydrolase is located on the inner envelope membrane (Block *et al.*, 1983a), again aiding the retention of the diacylglycerol. The diacylglycerol: UDP galac-tosyltransferase which uses the substrate is also located in the inner envelope membrane of spinach (Block *et al.*, 1983b). By contrast, the galactosyltransferase in pea chloroplasts, which uses externally-generated diacylglycerol, appropriately is located in the outer envelope (Cline & Keegstra, 1983). Further characteristics of the diacylglycerol: UDP-galac-tose galactosyltransferases have been well described (Joyard & Douce, 1987).

Less straightforward is our conception of digalactosyldiacylglycerol formation. What is clear is that this lipid is also formed in the plastid. Two enzymes have been invoked (Fig. 4), one using UDP-galactose as the source of the second galactose moiety while a second pathway uses galactolipid: galactolipid galactosyltransfer. Both activities have been assayed in plastid envelopes (see Joyard & Douce, 1987). Douce and co-workers believe that the galactolipid transferase is involved in catabolism rather than anabolism but as pointed out by Harwood (1989) the galactosylation of monogalactosyldiacylglycerol has been reported to occur in two phases – each of which could be catalysed by a different enzyme.

Galactosylglyceride biosynthesis

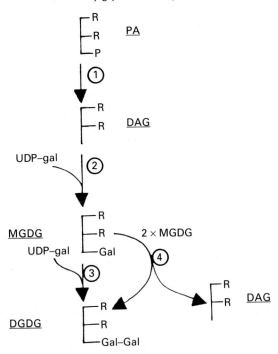

Fig. 4. Synthetic pathways for galactosylglyceride synthesis, 1=Phosphatidate phosphohydrolase (PAPase); 2=UDP-galactose (UDP-galactose:diacylglycerol galactosyltransferase); 3=UDP-galactose: monogalactosyldiacylglycerol (MGDG) galactosyltransferase; 4=MGDG: MGDG galactosyltransferase.

Phosphoglyceride synthesis

The distribution of phosphoglycerides in plant cells is quite distinct. Typically, phosphatidylglycerol is the main or only phospholipid in thylakoids (Gounaris *et al.*, 1986) while diphosphatidylglycerol is confined to the inner mitochondrial membrane (Bligny & Douce, 1980). As befits such a distribution, phosphatidylglycerol synthesis has been demonstrated in isolated plastids (Andrews & Mudd, 1985) while diphosphatidylglycerol formation has been shown indirectly in mitochondria (see Moore, 1982).

Both phosphatidylinositol (PI) and phosphatidylglycerol are formed from CDP-diacylglycerol whereas the nitrogen-containing phosphatidylcholine (PC) and phosphatidylethanolamine (PE) utilize the CDP-base pathway, primarily. Not surprisingly, the later stages of the formation of

Table 4. *Demonstrated subcellular location of enzymes involved in the synthesis of phosphoglycerides*

Enzyme	Location
Glycerol 3-phosphate acyltransferase	Plastid stroma Endoplasmic reticulum
Lysophosphatidate acyltransferase	Plastid membranes Endoplasmic reticulum
Phosphatidate phosphohydrolase	Plastid envelope Endoplasmic reticulum
Choline/ethanolamine kinases	Cytosol
Cholinephosphate cytidylyltransferase	Cytosol/endoplasmic reticulum
DAG:CDP-choline choline phosphotransferase	Endoplasmic reticulum
Phosphatidylethanolamine methylation	Endoplasmic reticulum Inner mitochondrial membrane
Phosphatidylinositol synthetase	Endoplasmic reticulum Inner mitochondrial membrane Plastid envelope
Phosphatidate: CTP cytidylyltransferase	Endoplasmic reticulum Inner mitochondrial membrane Plastid envelope

PC, PE and PI have all been localized in the endoplasmic reticulum. Interestingly, the phosphorylated derivatives of PI which have recently been demonstrated in plants (see Harwood, 1989) seem to be formed in the plasma membrane – an analogous situation to mammals. The demonstrated subcellular location of a number of phospholipid-metabolizing enzymes is listed in Table 4.

Possible roles of lipid exchange proteins

It will have become clear by now that lipids have to be moved around the plant cell at a number of stages in metabolism. Depending on the plant species, these movements may occur when:

(a) fatty acyl-CoAs leave the plastid
(b) the diacylglycerol backbone is transferred from the endoplasmic reticulum back to the plastid in '18:3'-plants

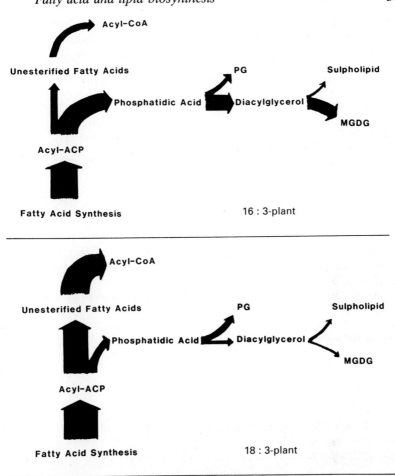

Fig. 5. Key differences in lipid metabolism between '16:3' and '18:3'-plants. PG=phosphatidylglycerol; MGDG=monogalactosyldiacylglycerol. The thickness of the reaction lines denotes their relative importance (Re-drawn from Joyard & Douce, 1987).

(c) phosphoglycerides have to be moved from their site of synthesis (endoplasmic reticulum) to other membranes
(d) glycosylglycerides synthesized in the envelope have to enter the plastid thylakoids.

Virtually nothing is known about the mechanism of their movement *in vivo*. Nevertheless, a number of phospholipid (and one galactolipid) exchange proteins have been isolated and studied. The details of these proteins and their possible functions are reviewed by Kader *et al.* (1984).

Further remarks on '16:3' and '18:3'-plants

Key differences between the metabolism of lipids in '16:3'-plants and '18:3'-plants are shown in Fig. 5. Thus, from the time that acyl groups are preferentially hydrolysed from acyl-ACPs rather than being used for acyl transferase, their fate in '18:3' plants has been determined. The activities and location of the various other enzymes involved have an enabling rather than controlling role. Therefore, the experimental observations by Guerra *et al.* (1986) where ACP isoforms influence the activity of acyl-ACP hydrolase and acyltransferase is particularly interesting. They offer the possibility of using molecular biology to increase carbon fluxes through particular pathways. However, it is also possible that by so doing, another enzyme reaction will become controlling, or, to put it another way:

'the solution to a problem changes the problem' (Peer's Law).

References

Agrawal, V. P. & Stumpf, P. K. (1985). Characterisation and solubilization of an acyl chain elongation system in microsomes of leek epidermal cells. *Arch. Biochem. Biophys.* **240**, 154–65.

Andrews, J. & Heinz, E. (1987). Desaturation of newly synthesized monogalactosyldiacylglycerol in spinach chloroplasts. *J. Plant Physiol.* **131**, 75–90.

Andrews, J. & Mudd, J. B. (1985). Phosphatidylglycerol synthesis in pea chloroplasts. *Plant Physiol.* **79**, 259–65.

Appleby, R. S., Gurr, M. I. & Nichols, B. W. (1974). Factors controlling the biosynthesis of fatty acids and acyl lipids in subcellular organelles in maturing *Crambe abyssinica* seeds. *Eur. J. Biochem.* **48**, 209–16.

Bligny, R. & Douce, R. (1980). A precise localisation of cardiolipin in plant cells. *Biochim. Biophys. Acta* **617**, 254–63.

Block, M. A., Dorne, A.-J., Joyard. J. & Douce, R. (1983a). The phosphatidic acid phosphatase of the chloroplast envelope is located on the inner envelope membrane. *FEBS Lett.* **164**, 111–15.

Block, M. A., Dorne, A.-J., Joyard, J. & Douce, R. (1983b). Preparation and characterization of membrane fractions enriched in outer and inner envelope membranes from spinach chloroplasts. *J. Biol. Chem.* **258**, 13281–6.

Cao, Y.-Z. & Huang, A. H. C. (1986). Diacylglycerol acyltransferase in maturing oil seeds of maize and other species. *Plant Physiol.* **82**, 813–20.

Cassagne, C., Lessire, R., Bessoule, J. & Moreau, P. (1987). Plant elongases. In *The Metabolism, Structure and Function of Plant Lipids*, ed. P. K. Stumpf, J. B. Mudd & W. D. Nes, pp. 481–8. New York: Plenum Press.

Caughey, I. & Kekwick, R. G. O. (1982). The characteristics of some components of the fatty acid synthetase system in the plastids from the mesocarp of avocado fruit. *Eur. J. Biochem.* **123**, 553–61.

Cline, K. & Keegstra, K. (1983). Galactosyltransferases involved in galactolipid biosynthesis are located in the outer membrane of pea chloroplast envelopes. *Plant Physiol.* **71**, 366–72.

Frentzen, M., Heinz, E., McKeon, T. M. & Stumpf, P. K. (1983). Specificities and selectivities of glycerol-3-phosphate acyltransferase and monoacylglycerol-3-phosphate acyltransferase from pea and spinach chloroplasts. *Eur. J. Biochem.* **129**, 629–36.

Gardiner, S. E. & Roughan, P. G. (1983). Relationship between fatty acid composition of diacylgalactosylglycerol and turnover of chloroplast phosphatidate. *Biochem. J.* **210**, 949–52.

Gounaris, K., Barber, J. & Harwood, J. L. (1986). The thylakoid membranes of higher plant chloroplasts. *Biochem. J.* **237**, 313–26.

Griffiths, G., Stobart, A. K. & Stymne, S. (1985). The acylation of *sn*-glycerol 3-phosphate and the metabolism of phosphatidate in microsomal preparations from the developing cotyledons of safflower seed. *Biochem. J.* **230**, 379–88.

Guerra, D. J., Ohlrogge, J. B. & Frentzen, M. (1986). Activity of acyl carrier protein isoforms in reactions of plant fatty acid metabolism. *Plant Physiol.* **82**, 448–53.

Gunstone, F. G., Harwood, J. L. & Padley, F. B., (Eds.) (1986). *The Lipid Handbook*. London: Chapman and Hall.

Gurr, M. I. (1980). The biosynthesis of triacylglycerols. In *The Biochemistry of Plants*, vol. 4, ed. P. K. Stumpf & E. E. Conn, pp. 203–48. New York: Academic Press.

Hardie, D. G., Carling, D. & Slim, A. T. R. (1989). *Trends Biochem. Sci.* **14**, 20–23.

Harwood, J. L. (1988). Fatty acid metabolism. *Ann. Rev. Plant Physiol.* **39**, 101–38.

Harwood, J. L. (1989). Lipid metabolism in plants. *C.R.C. Critical Revs. in Plant Sci.* **8**, 1–43.

Harwood, J. L. & Price-Jones, M. J. (1987). Phosphatidate phosphohydrolase of plants and microorganisms. In *Phosphatidate Phosphohydrolase*, vol. 2, ed. D. N. Brindley, pp. 1–37. Boca Raton: CRC Press.

Hoj, P. B. & Mikkelson, J. D. (1982). Partial separation of individual enzyme activities of the fatty acid synthetase from barley chloroplasts. *Carlsberg Res. Commun.* **47**, 119–41.

Ichihara, K. (1984). sn-Glycerol-3-phosphate acyltransferase in a particulate fraction from maturing safflower seeds. *Arch. Biochem. Biophys.* **232**, 685–98.

Jackowski, S. & Rock, C. O. (1987). Acetoacetyl acyl carrier protein synthase, a potential regulator of fatty acid biosynthesis in bacteria. *J. Biol. Chem.* **262**, 7927–31.

Jaworski, J. G., Clough, R. C. & Barnam, S. R. (1989). A cerulenin insensitive short chain 3-ketoacyl-acyl carrier protein synthase in *Spinacia olearacea* leaves. *Plant Physiol.* **90**, 40–4.

Jones, A. V. M. & Harwood, J. L. (1980). Desaturation of linoleic acid from exogenous lipids by isolated chloroplasts. *Biochem. J.* **190**, 851–4.

Joyard, R. & Douce, R. J. (1987). Galactolipid synthesis. In *The Biochemistry of Plants*, vol. 9, ed. P. K. Stumpf & E. E. Conn, pp. 215–74. New York: Academic Press.

Joyard, J. & Stumpf, P. K. (1981). Synthesis of long-chain acyl-CoA in chloroplast envelope membranes. *Plant Physiol.* **67**, 250–6.

Joyard, J., Chuzel, M. & Douce, R. (1979). Is the chloroplast envelope a site of galactolipid synthesis? Yes. In *Advances in the Biochemistry and Physiology of Plant Lipids*, ed. L. A. Appelqvist & C. Liljenberg, pp. 181–6. Amsterdam: Elsevier.

Kader, J., Douady, D., Grosbois, M., Guerbetteo, F. & Vergnolle, C. (1984). The role of intracellular lipid movements in membrane biogenesis. In *Structure, Function and Metabolism of Plant Lipids*, ed. P. A. Siegenthaler & W. Eichenberger, pp. 283–90. Amsterdam: Elsevier.

Kobek, K., Focke, M., Lichtenthaler, H. K., Retzlaff, G. & Wurzer, B. (1988). Inhibition of fatty acid biosynthesis in isolated chloroplasts by cycloxydim and other cyclohexane-1,3-diones. *Physiologia Plant* **72**, 492–8.

Lessire, R., Bessoule, J. J. & Cassagne, C. (1985). Solubilisation of C_{18}-CoA and C_{20}-CoA elongases from *Allium porrum* L. epidermal cell microsomes. *FEBS Lett.* **187**, 314–20.

McHenry, L. & Fritz, P. J. (1987). Cocoa butter biosynthesis: effect of temperature on *Theobroma cacao* acyltransferases. *J. Am. Oil Chem. Soc.* **64**, 1012–15.

McKeon, T. M. & Stumpf, P. K. (1982). Purification and characterisation of the stearoyl-ACP desaturase and acyl-ACP thioesterase from maturing seeds of safflower. *J. Biol. Chem.* **257**, 12141–47.

Moore, T. S. (1982). Phospholipid biosynthesis. *Ann. Rev. Plant Physiol.* **33**, 235–59.

Murata, N., & Nishida, I. (1987). Lipids of blue-green algae (Cyanobacteria). In *The Biochemistry of Plants*, vol. 9, ed. P. K. Stumpf & E. E. Conn, pp. 315–47. New York: Academic Press.

Nikolau, B. & Hawke, J. C. (1984). Purification and characterisation of maize leaf acetyl-coenzyme A carboxylase. *Arch. Biochem. Biophys.* **228**, 86–96.

Nishida, I., Frentzen, M., Ishizaka, O. & Murata, N. (1987). Purification of isomeric forms of acyl-[acyl-carrier-protein]: glycerol-3-phosphate acyltransferase from greening squash cotyledons. *Plant Cell Physiol.* **28**, 1071–9.

Ohlrogge, J. C. (1987). Biochemistry of plant acyl carrier protein. In *The Biochemistry of Plants*, vol. 9, ed. P. K. Stumpf & E. E. Conn, pp. 137–57. New York: Academic Press.

Ohlrogge, J. B., Kuhn, D. N. & Stumpf, P. K. (1979). Subcellular localisation of acyl carrier protein in leaf protoplasts of *Spinacia oleracea*. *Proc. Natn. Acad. Sci., USA* **76**, 1194–8.

Ohnishi, J. I. & Yamada, M. (1982). Glycerolipid synthesis in *Avena* leaves during greening of etiolated seedlings. III. Synthesis of linolenoyl-monogalactosyldiacylglycerol from liposomal linoleoyl-phosphatidylcholine by *Avena* plastids in the presence of phosphatidylcholine – exchange protein. *Plant Cell Physiol.* **23**, 767–73.

Pollard, M. R. & Singh, S. S. (1987). Fatty acid synthesis in developing oil seeds. In *The Metabolism, Structure and Function of Plant Lipids*, ed. P. K. Stumpf, J. B. Mudd & W. D. Nes, pp. 455–63. New York: Plenum Press.

Pugh, E. L. & Kates, M. (1973). Desaturation of phosphatidylcholine and phosphatidylethanolamine by a microsomal enzyme system from *Candida lipolytica*. *Biochim. Biophys. Acta* **316**, 305–16.

Roughan, P. G. (1987). On the control of fatty acid compositions of plant glycerolipids. In *The Metabolism, Structure and Function of Plant Lipids*, ed. P. K. Stumpf, J. B. Mudd & W. D. Nes, pp. 247–54. New York: Plenum Press.

Roughan, P. G., Mudd, J. B., McManus, T. T. & Slack, C. R. (1979). Linoleate and α-linolenate synthesis by isolated spinach chloroplasts. *Biochem. J.* **184**, 571–4.

Sato, N. & Murata, N. (1982). Lipid biosynthesis in the blue green-alga, *Anabaena variabilis*. II. Fatty acids and lipid molecular species. *Biochim. Biophys. Acta* **710**, 279–89.

Sato, N., Seyama, Y. & Murata, N. (1986). Lipid-linked desaturation of palmitic acid on monogalactosyldiacylglycerol in the blue-green alga *Anabaena variabilis* studied *in vivo*. *Plant Cell Physiol.* **27**, 819–35.

Shimakata, T. & Stumpf, P. K. (1983). The purification and function of acetyl coenzyme A: acyl carrier protein transacylase. *J. Biol. Chem.* **258**, 3592–8.

Shine, W. E., Mancha, M. & Stumpf, P. K. (1976). The function of acyl thioesterases in the metabolism of acyl-CoAs and acyl-ACPs. *Arch. Biochem. Biophys.* **172**, 110–16.

Slabas, A. R. & Hellyer, A. (1985). Rapid purification of a high molecular weight subunit polypeptide form of rape seed acetyl-CoA carboxylase. *Plant Sci.* **39**, 177–82.

Slabas, A. R., Harding, J., Hellyer, A., Sidebottom, C., Gwynne H., Kessell, R. & Tombs, M. P. (1984). Enzymology of plant fatty acid biosynthesis. In *Structure, Function and Metabolism of Plant Lipids*, ed. P. A. Siegenthaler & Eichenberger, W., pp. 3–10. Amsterdam: Elsevier.

Stumpf, P. K. (1987). The biosynthesis of saturated fatty acids. In *The Biochemistry of Plants*, vol. 9, ed. P. K. Stumpf & E. E. Conn, pp. 121–36. New York: Academic Press.

Stymne, S. & Appelqvist, L.-A. (1978). The biosynthesis of linoleate from oleoyl-CoA via oleoyl-phosphatidylcholine in microsomes from developing safflower seeds. *Eur. J. Biochem.* **90**, 223–9.

Stymne, S. & Stobart, A. K. (1987). Triacylglycerol biosynthesis. In *The Biochemistry of Plants*, vol. 9, ed. P. K. Stumpf & E. E. Conn, pp. 175–214. New York: Academic Press.

Stymne, S., Green, A. G. & Tonnet, M. L. (1989). Lipid synthesis in developing cotyledons of linolenic acid deficient mutants of linseed. In *Biological Role of Plant Lipids*, eds. P. A. Biacs, K. Gruiz & T. Kremmer. Budapest and London: Plenum Press and Akademiai Kiado, pp. 147–50.

Thompson, G. A., Roughan, P. G., Browse, J. A., Slack, C. R. & Gardiner, S. E. (1986). Spinach leaves desaturate exogenous [^{14}C]palmitate to hexadecatrienoate. *Plant Physiol.* **82**, 357–62.

Turnham, E. & Northcote, D. N. (1983). Changes in the activity of acetyl-CoA carboxylase during rapseseed formation. *Biochem. J.* **212**, 223–9.

Walker, K. A. & Harwood, J. L. (1985). Localisation of chloroplastic fatty acid synthesis *de novo* in the stroma. *Biochem. J.* **226**, 551–6.

Walker, K. A. & Harwood, J. L. (1986). Evidence for separate elongation enzymes for very long chain fatty acid synthesis in potato. *Biochem. J.* **237**, 41–6.

Walker, K. A., Ridley, S. M. & Harwood, J. L. (1988). Effects of the selective herbicide fluazifop on fatty acid synthesis in pea and barley. *Biochem. J.* **254**, 811–17.

Walker, K. A., Ridley, S. M., Lewis, T. & Harwood, J. L. (1989). Action of aryloxy-phenoxy carboxylic acids on lipid metabolism. *Revs. Weed Sci.*, **4**, 71–84.

Weaire, P. J. & Kekwick, R. G. O. (1975). The fractionation of the fatty acid synthetase activities of avocado mesocarp plastids. *Biochem. J.*, **146**, 439–445.

A. H. C. HUANG, R. QU, Y. K. LAI,
C. RATNAYAKE, K. L. CHAN, G. W. KUROKI,
K. C. OO, and Y. Z. CAO

Structure, synthesis and degradation of oil bodies in maize

Introduction

Most seeds contain storage oils in the form of triacylglycerols, which usually comprise 10–50% of the total seed dry weight (Gurr, 1980; Stymne & Stobart, 1987). This oil reserve is synthesized during seed maturation, and is rapidly mobilized to provide energy and carbon skeletons for the growth of the embryo in postgermination. The triacylglycerols are densely packed in subcellular organelles called oil bodies (lipid bodies, oleosomes, spherosomes). The spherical oil body is about 0·5–1 μm in diameter, and is surrounded by a 'half-unit' membrane of one layer of phospholipids about 3 nm thickness (Yatsu & Jacks, 1972). The acyl moieties of the membrane phospholipids orient themselves toward the matrix so that they can form hydrophobic interactions with the internal triacylglycerols.

The ontogeny of the oil bodies during seed maturation is still unclear. It is generally believed that fatty acids are synthesized in the plastids (Gurr, 1980; Stymne & Stobart, 1987). The subsequent formation of mono-, di-, and triacylglycerols from activated fatty acids occurs in the microsomes, which represent mainly the endoplasmic reticulum (ER). The mechanism of transport of fatty acid from the plastids to the ER is unknown. The origin of the oil body membrane has been investigated recently. The membrane phospholipids are generally assumed to be synthesized in the ER. It has been suggested that the newly synthesized triacylglycerols in the ER are sequestered between the two phospholipid layers of the membrane at a particular region, such that a budding vesicle of triacylglycerols surrounded by a layer of phospholipid is formed (Schwarzenbach, 1971; Wanner *et al.*, 1981; Huang *et al.*, 1987). The vesicle is then detached from the ER to become an oil body. An alternative postulation states that the oil bodies arise directly in the cytoplasm by condensation of ER-derived triacylglycerol molecules together with, or followed by, the formation of a surrounding membrane (Stymne & Stobart, 1987; Bergfeld *et*

43

al., 1978; Herman, 1987; Murphy *et al.*, 1989). The fate of the oil body membrane after lipolysis in postgermination has only been studied recently.

Diacylglycerol acyltransferase is localized in the rough endoplasmic reticulum

Diacylglycerol acyltransferase (EC 2·3·1·20) catalyses the final step in the synthesis of triacylglycerols in oil seed (Gurr, 1980; Stymne & Stobart, 1987). It is the only known enzyme unique to the long biosynthetic pathway of triacylglycerols, since the diacylglycerol produced in the preceding step could also be used to produce phospholipids or galactolipids. The enzyme, alone or in conjunction with the enzymes in preceding metabolic steps, has been detected in the microsomes from the maturing seeds of several species.

We have carried out subcellular fractionation of the total extracts of maize scutellum and castor bean endosperm in sucrose density gradients (Cao & Huang, 1986). By comparing the migration of the enzyme between rate and equilibrium centrifugation in the presence and absence of magnesium ions in the preparative media, the enzyme has been shown to be associated with the rough ER.

Oil body membrane contains unique proteins, the oleosins

In maize, the isolated oil bodies contain 97–98% by weight triacylglycerols, 1% phospholipids, and 1·7% proteins (our unpublished data). The acyl moieties in the triacylglycerols include about 60% linoleic acid, 25% oleic acid, and 10% stearic acid. There are no unusual phospholipid components, and phosphatidylethanolamine represents about 14%, phosphatidylserine/phosphatidylinositol 16%, phosphatidylcholine 45%, phosphatidic acid 5%, and phosphatidylglycerol 5%.

The proteins of the maize oil bodies can be resolved into several major protein bands by SDS-polyacrylamide gel electrophoresis (Qu *et al.*, 1986; Fig. 1). There are three bands of low M_rs (19,500, 18,000, and 16,500, termed L1, L2, and L3, respectively) and one band of higher M_r (40,000, termed H). The L proteins are unique to the oil bodies, whereas the H protein is also present in the microsomes. The presence of H protein in the oil bodies does not appear to be the result of contamination of the oil body fraction by microsomes, as many microsomal protein bands of higher amounts are absent in the oil body fraction. Immunocytochemistry shows that L3 is restricted to the membrane of the oil bodies (Fernandez *et al.*, 1988; Fig. 2). The L proteins have alkaline pI values, and are hydrophobic integral proteins, as shown by their

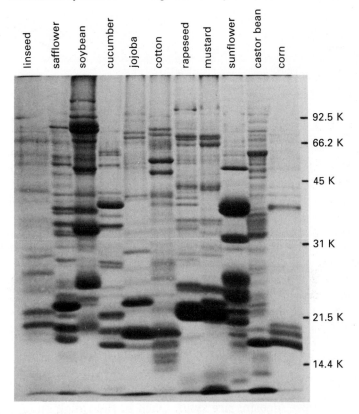

Fig. 1. SDS polyacrylamide gel electrophoresis of oil body proteins isolated from seeds of various plant species. Positions of M_r markers are shown on the right. In the lane containing maize (corn) proteins, the three bands of M_r 19,500, 18,000, and 16,500 are termed L1, L2, and L3, respectively, and the band of M_r 40,000 is called H (Qu *et al.*, 1986).

resistance to solubilization from the oil bodies after repeated washing, amino acid composition, and partitioning in a Triton X-114 system. The L proteins can be purified to apparent homogeneity by preparative SDS-polyacrylamide gel electrophoresis. We have studied L2 and L3 most extensively, and found that they share very similar chemical properties.

We hereby coin the general term 'oleosins' to describe collectively the L proteins in maize oil bodies as well as similar proteins in the oil bodies of other seed species. As to be described, oleosins with similar properties are ubiquitous in the oil bodies of diverse species.

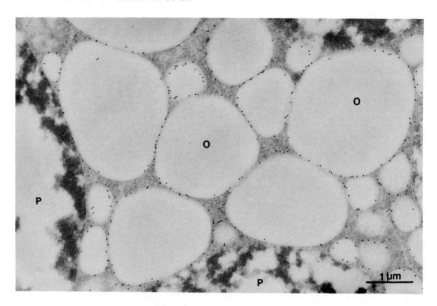

Fig. 2. Electron micrograph of postgerminated (0·5 day) maize kernel cell immunolabelled with antibodies directed against the L3 protein. O, oil bodies; P, protein bodies (Fernandez *et al.*, 1988).

Maize oleosins are synthesized in the rough endoplasmic reticulum without apparent processing

Labelling *in vivo* with 35-S methionine and translation *in vitro* using extracted RNA in a wheatgerm system reveal that the maize oleosins are synthesized during seed maturation and not in postgermination (Qu *et al.*, 1986). The proteins synthesized *in vivo* and *in vitro* have no appreciable differences in their mobilities in two-dimensional gel electrophoresis (isoelectric focusing and molecular sieving). L3 is synthesized predominantly, if not exclusively, by RNA derived from bound polyribosomes and not from free polyribosomes. The lack of a cleavable signal peptide in oleosins during synthesis is not surprising as the proteins are not for export. In addition, their hydrophobicity can provide facile recognition for the hydrophobic region of the budding oil bodies.

Further evidence for the oleosins being synthesized in the rough ER is our success in isolating cDNA clones of L2 (in preparation) and L3 (Vance & Huang, 1987) using size-select mRNA from bound polyribosomes. We estimate that at least 50% of the size-select mRNA from bound polyribosomes represent those for the synthesis of L2 and L3. This high percentage reflects the suitability of the size-selection as well as the high amounts of L proteins and their mRNA in the cell.

Oleosin has distinct structural features which enable it to reside stably on the oil bodies

We have obtained and sequenced cDNA clones of L2 (in preparation) and L3 (Vance & Huang, 1987). The authenticity of these clones has been confirmed by the complete identity between the amino acid sequences deduced from the nucleotide sequences and those obtained by N-terminal sequencing of the isolated peptides after protease digestion (14 residues of proteinase Lys-C-digested L2, and 15 residues of proteinase V8-digested L3; data unpublished). The amino acid sequences deduced from the nucleotide sequences were analysed for their secondary structures using the program by Chou & Fasman (1978) and the prediction of Engleman & Steitz (1981). Both L2 and L3 have very similar secondary structures, each consisting of three structural domains (Fig. 3). The N-

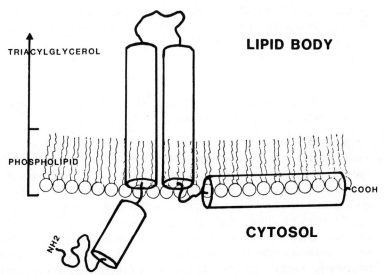

Fig. 3. Proposed model for the topology of L3 protein on the oil body. Cylinders are used to depict α-helices. The phospholipid and triacylglycerol moieties of the oil body are demarcated. The protein can be viewed as consisting of three structural domains: the N-terminal hydrophilic α-helix, the central hydrophobic hairpin α-helix, and the C-terminal amphipathic α-helix (Vance & Huang, 1987).

terminal segment contains a hydrophilic domain which should expose itself freely into the cytosol. The central segment of the protein forms a hydrophobic hairpin α-helical structure. In the middle of this segment, there are three proline residues which allow the breaking and bending of the helical structure. The hairpin structure is predicted to penetrate into the hydrophobic core of the oil bodies. The C-terminal segment is an amphipathic α-helix which contains basic amino acid residues at the amphipathic junction. The structure should be able to reside stably on the amphipathic surface of an oil body, and the junctional basic amino acid residues can interact with the acidic phosphate groups of the monolayer phospholipids.

Although both L2 and L3 contain the same three major structural domains, a high degree of similarity between their amino acid sequences occurs only in the central domain of the hydrophobic hairpin α-helix. The similarity is 75% when substitution of similar amino acid residues is also included. Of the 17 amino acid residues in the hairpin turn (in the middle) of this domain, the similarity is 100%, and 15 of the 17 amino acid residues are identical. On the other hand, the amino acid sequences of the N-terminal hydrophilic domain of L2 and L3 share a low degree of similarity (36%). Also, even though both the third domains of L2 and L3 contain clearcut amphipathic α-helices with basic amino acids occurring at the amphipathic junctions, their amino acid sequences have a low degree of similarity (38%).

The above comparison between the structures of L2 and L3 have several implications. First, the N-terminal hydrophilic domain and the amphipathic α-helix domain have much less structural/functional constraints (relative to the central hairpin region) such that extensive substitutions of the amino acids along the polypeptides are possible. Second, these two structural domains, especially the N-terminal hydrophilic domain, are exposed to the aqueous phase in an antigen-adjuvent emulsion, and therefore represent the antigenic sites of the proteins. The central domain of the hairpin structure is highly hydrophobic and should be buried in an antigen-adjuvant emulsion, and thus is much less antigenic. These differences in the antigenicity explain our observation that the antibodies prepared against L2 and L3 do not cross-react with their respective antigens. Third, amphipathic α-helices are common features of many mammalian apolipoproteins (Driscoll & Getz, 1986). The amino acid sequences between these mammalian amphipathic α-helices and the oleosin amphipathic helices do not share a high degree of similarity. Therefore, there appear to be many possible makeups of amphiphathic α-helices which can satisfy the structural constraints for an association with the surface of a monolayer of phospholipids surrounding a core of neutral lipids. Fourth, the close similarity between the amino acid sequences of the hairpin α-helices of L2 and L3 reflects a severe structural/functional constraint imposed by the hydrophobic core of the oil body. This implication is supported by our observation that the hairpin regions in the oleosins of several diverse seed species also have highly similar amino acid sequences (next section). Fifth, although the L proteins are synthesized in the rough ER, they do not have an N-terminal signal sequence that is processed co- or post-translationally. The signal residing in the L proteins, which is required for subcellular protein trafficking, may be provided by the very hydrophobic α-helix hairpin domain. This domain occurs in the L proteins as well as in the oleosins of other seed species (next section). In L proteins, each of the α-helics of the hairpin has 30–33 amino acid residues, which constitutes an α-helix about 5 nm long. This length, plus an additional but unknown value due to the middle polypeptide turn, makes the hairpin longer than the thickness of the acyl portion (4·5 nm) of a bilayer phospholipid membrane. Therefore, the hairpin structure will be stable on the monolayer phospholipid surface of an oil body, but not on a bilayer phospholipid of a normal cell membrane. We propose that the hairpin domain provides a recognition signal for oleosins in subcellular protein trafficking as well as a strong anchorage for the protein on the oil body surface. This proposal may explain why hydrophobic hairpin α-helices containing highly similar amino acid sequences are present in oleosins from several seed species studied. The hairpin domain would make the protein unstable in a normal bilayer phospholipid membrane, and this instability may explain the selective removal of the maize L proteins from the membrane after lipolysis in postgermination (latter part of this article).

Oleosins with characteristics similar to those in maize are present in diverse seed species

We have studied the proteins in oil bodies isolated from many seed species, including linseed, safflower, soybean, cucumber, jojoba, cotton, rapeseed, mustard, sunflower, and castor bean (Qu *et al.*, 1986; Fig. 1) as well as peanut, palm, wheat, and barley (unpublished). These oil body proteins can be resolved into a few distinct protein bands by SDS-polyacrylamide gel electrophoresis, and the protein band patterns vary greatly among diverse species. Those from closely related species are very similar (e.g. rapeseed and mustard of the *Brassica* genus). A major feature common to the oil body proteins of all the seed species is that they all possess abundant proteins of low M_rs (15–30,000). Some of the minor protein bands may represent contaminants in the oil body fractions. The oil body proteins of soybean (24 kDa) and rapeseed (19 kDa) (Fig. 1) have subsequently been studied by other laboratories (Herman, 1987; Murphy *et al.*, 1989). These latter workers have shown that the proteins are also hydrophobic and localized in the oil bodies, and that the soybean 24 kDa protein is synthesized without apparent co- or post-translational processing.

We have subjected the abundant proteins of oil bodies from diverse species to protease digestion, and analysed the resulting protein fragments for their N-terminal amino acid sequences. A small segment of the rapeseed oil body protein has also been sequenced similarly (Murphy & Au, 1989). Although the sequencing of the protein fragments is far from completion, the findings reiterate our interpretation of the structures of L2 and L3 from maize. All studied major oil body proteins contain at least a central domain which possesses a hydrophobic hairpin α-helix having very similar amino acid sequences. Again, this common region is considered less antigenic, and thus antibodies prepared against the maize L proteins react weakly (e.g. barley and wheat) or not at all (e.g. rapeseed) with the oil body proteins from other species. We hereby consider these oil body proteins to be oleosins because of their similarities in structure, abundance, and subcellular localization. It is likely that they are encoded by genes belonging to the same gene family having evolved from a common ancestor gene.

Using the Sequence Analysis Software Package of the Genetics Computer Group of the University of Wisconsin, we did a search for proteins which have amino acid sequences similar to those of the maize L proteins. The proteins topping the list have similarities only at very restricted regions along the whole amino acid sequences, and are not considered of statistical significance. The mammalian apolipoproteins, in spite of their associating with neutral lipid particles and having substantial amphipathic α-helical structure and hydrophobic regions, do not share significant similarities in amino acid sequences with the maize L proteins.

Oleosins in maize are encoded by a small gene family

The cDNA sequences of L2 and L3 share a high degree of similarities at the stretch encoding the central domain of the proteins; the degree of similarity is 71% within this stretch of 225 nucleotides.

Genomic clones of maize L2 and L3 have been obtained. There are no introns in either gene. Restriction enzyme analyses of genomic DNA in northern blotting show that each gene is represented by a single copy or a few copies per haploid genome. L2 and L3 have been mapped by restriction fragment length polymorphism linkage analyses at single loci on chromosome 5 and chromosome 2, respectively (courtesy of T. Helent-jaris of Native Plant Institute).

The expression of the L2 (in preparation) and L3 (Vance & Huang, 1988) genes are coordinate and tissue-specific, and are under temporal, developmental, and hormonal controls. The genes are expressed only in the embryo and the aleurone layer; both tissues are known to contain oil bodies. In addition, their mRNAs increase during seed maturation, but are reduced to a low level in mature seed. When the mature seed is allowed to imbibe in the presence of exogenous abscisic acid, the expression of L2 and L3 is extended and enhanced concomitant with the retardation of germination. These controls occur at least at the transcription level.

In postgerminative seedling growth, lipase activity increases concomitant with the decrease in storage oil

In postgerminative seedling growth, lipase (EC 3·1·1·3) hydrolyses the storage triacylglycerols (Huang, 1987). Most of the products (fatty acid and glycerol) are used for gluconeogenesis to support the growth of the embryonic axis. In most seed species, lipase activities are absent in the ungerminated seed and increase in postgermination. The only well-documented exception to this developmental pattern is castor bean which has active lipase in the ungerminated seeds. The lipases from several representative seed species have been shown to be relatively specific for triacylglycerols which contain the major fatty acid components of the storage triacylglycerols in the same seeds.

Lipase is localized in oil bodies in maize

After subcellular fractionation of the storage tissues of many germinated seeds, the lipase activity is present either in the soluble fraction or associated with the membranes of the oil bodies (Huang, 1987). In seeds where the lipase is found in the soluble fractions, the enzymes *in vivo* will still have to come in contact with the membrane of the oil bodies during catalysis. In those seeds where the lipase is associated with the membrane of the oil bodies, the enzyme may be loosely (e.g. rapeseed and mustard seed) or tightly (e.g. castor bean and maize) associated with the organelles.

The classical studies of castor bean lipase have been well-documented (Ory, 1969). The only plant lipase that has been purified to homogeneity and its biosynthesis studied, is the enzyme from maize. In the scutella of maize, lipase activity is absent in ungerminated seeds and increases in postgermination (Lin & Huang, 1984). At the peak stage of lipolysis, about 60% of the lipase activity can be recovered in the oil body fraction after flotation centrifugation. The lipase is tightly bound to the oil bodies, and resists solubilization by repeated washing with buffers or NaCl solution. The lipase has been purified 272-fold to apparent homogeneity. The enzyme in sodium deoxycholate has an approximate M_r of 27,000 by sucrose gradient centrifugation and a M_r of 65,000 by SDS-polyacrylamide gel electrophoresis. The amino acid composition as well as a biphasic partition using Triton-X 114 reveals the enzyme to be a hydrophobic protein.

Maize lipid is synthesized in free polyribosomes without apparent co- or post-translational modification

The biosynthesis of the maize lipase has been pursued using monospecific rabbit antibodies raised against the purified lipase (Wang & Huang, 1987). Using an *in vitro* protein synthesis system, the mRNA for the lipase can be detected in germinated but not maturing seeds. The *in vitro* and *in vivo* synthesized lipase exhibit the same M_r (65,000) by SDS-polyacrylamide gel electrophoresis, and thus there is no apparent co- or post-translational processing of the lipase. The enzyme is synthesized by mRNA extracted from free and not bound polyribosomes. Apparently, after its synthesis, the lipase will attach itself specifically to the membrane of the oil bodies and not other cell organelles.

The maize lines, Illinois High Oil, Illinois Low Oil, and their F_1 generation, contain about 18, 0·5 and 10%, respectively, of kernel oils. Lipase activity which appears in postgermination is proportional to the oil content in each maize line (Wang *et al.*, 1984). This proportionality does not hold for catalase and isocitrate lyase, which are the same in the three maize lines. Thus, the selection for high oil content also produced high lipase activity. Since the lipase is synthesized *de novo* in postgermination, it is possible that the enzyme is synthesized or degraded in proportion to the available substrate or anchoring sites (i.e. oil body membrane).

Oil body membrane fuses with the vacuole membrane during or after lipolysis

The maize lipase activity starts to appear two days after imbibition, concomitant with the decrease in total lipids (Wang & Huang, 1987). The activity reaches a maximum at about day 5–6 (Fig. 4). The developmental profile of lipase activity at the initial stage of seedling growth parallels those of catalase and isocitrate lyase, two enzymes known to be involved

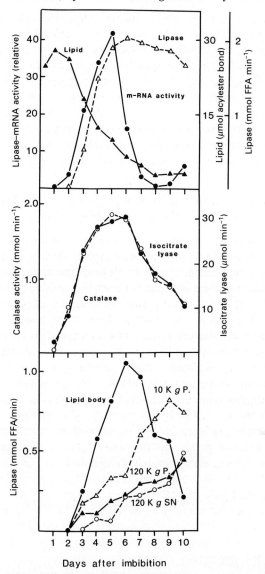

Fig. 4. Changes in total lipid, lipase activity, and lipase specific mRNA (upper panel), isocitrate lyase activity, and catalase activity (middle panel), and the distribution of lipase activities in different subcellular fractions (lower panel) in the scutella of maize kernel during seedling growth. All contents and activities are expressed on a per scutellum basis. FFA, free fatty acid; P, pellet; SN, supernatant (Wang & Huang, 1987).

in gluconeogenesis from oils. However, after reaching the maximum at day 5–6, lipase activity remains almost unaltered from day 5 to 10, whereas catalase and isocitrate lyase activities drop off rapidly. Apparently, either lipase is continuously being synthesized and turned over, or more likely the existing lipase is not being degraded. The latter prediction is supported by the complete disappearance of lipase specific mRNA after day 6 (Fig. 4).

At the peak stage of lipolysis (day 5–6), about 60% of the lipase activity can be recovered in the isolated membrane of the oil bodies, and the remaining activity (presumably representing nascent enzyme and enzyme derived from consumed oil bodies) is present in the $10\,000 \times g$ pellet, $120\,000 \times g$ pellet, and $120\,000 \times g$ supernatant. As the seedling grows beyond day 5, the proportion of lipase in the oil bodies decreases (concomitant with the decrease in triacylglycerol), whereas it increases in the $10\,000 \times g$ pellet, $120\,000 \times g$ pellet, and $120\,000 \times g$ supernatant. Since there is little change in the total lipase activity and no apparent synthesis of new lipase between days 5 and 10, it is likely that there is a transfer of the oil body membrane together with the lipase to a fragile compartment in the cell, presumably the vacuoles. Electron microscopic observation of the cells does show a physical connection between the membrane of degrading oil bodies and the membrane of enlarging cell vacuoles (Fig. 5). During the fusion, the oil body membrane should rearrange itself to form a double phospholipid layer.

In postgermination of maize seedling, oleosins disappear concomitant with the decrease in storage oil (Fernandez *et al.*, 1988). Thus, the preserved oil body membrane at the later stage of seedling growth apparently is devoid of the oleosins. As discussed earlier, it is possible that the oleosins are so hydrophobic that they become unstable in a phospholipid bilayer and are eliminated and degraded.

A model of oil body synthesis and degradation in maize seed

Our findings on maize oil bodies are consistent with a model of synthesis and degradation of the organelles during seed maturation and in postgermination (Fig. 6). During maturation, the oleosins are synthesized, without co- or post-translational modification, on polyribosomes on the rough ER. The rough ER also synthesizes triacylglycerols and membrane phospholipids of the oil bodies. These oil body components are sequestered at a localized region in the ER, such that an oil body is formed in which the triacylglycerol matrix is surrounded by a membrane of one phospholipid layer with embedded oleosins. In postgermination,

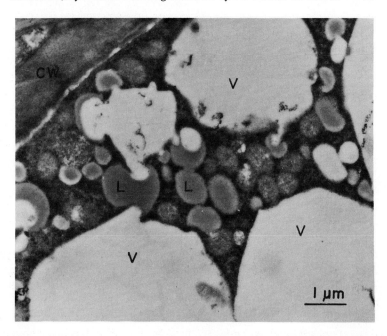

Fig. 5. Electron micrograph of part of a scutellum cell of day 6 seedling, showing degrading oil bodies (L), enlarging cell vacuoles (V), and cell wall (CW) (Wang & Huang, 1987).

lipase is synthesized on free polyribosomes. Without apparent co- or post-translational processing, the lipase moves to attach itself specifically to the membrane of the oil body. During or after lipolysis of the oil body, the organelle membrane together with the lipase fuses with the membrane of an enlarging cell vacuole. This fusion necessitates a rearrangement of the oil body membrane such that its monolayer of phospholipids is converted to a double layer of phospholipids of the vacuolar membrane. The membrane-attached lipase is retained whereas the oleosins are selectively removed and degraded. While our findings strongly support this model, they do not totally eliminate, but render unlikely, the alternative proposal that the oil body is synthesized as a naked triacylglycerol droplet and subsequently enclosed by a membrane. The original *in situ* electron microscopic observation supporting such a proposed biosynthetic model (Bergfeld *et al.*, 1978) has been disputed subsequently (Wanner *et al.*, 1981). In addition, the observed *in vitro* formation of naked oil droplets in microsomes supplied with acyl-CoA and glycerol (Stymne & Stobart, 1987) can be explained by the lack of a simultaneous

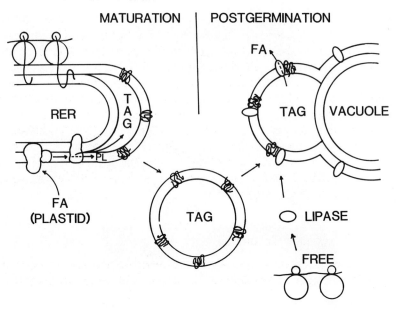

Fig. 6. A model for the synthesis and degradation of an oil body in maize scutellum during seed maturation and in postgermination. See text for explanation. RER, rough endoplasmic reticulum; FA, fatty acid; TAG, triacylglycerol; PL, phospholipid; FREE, free polyribosomes.

formation. of membrane phospholipids and proteins. Our model of oil body synthesis and degradation is based on our work with maize. Whether or not it also holds for other seed species remains to be seen.

Oleosins may serve the function of stabilizing the oil bodies and/or providing recognition signal for lipase anchorage

An intriguing question is the function of oleosins on the oil body membrane. Their abundance and the absence of a required enzymatic activity (other than lipase) in the oil bodies argue against their being enzymes. Seeds store triacylglycerols in many small oil bodies instead of one single large lipid droplet per cell as in mammalian white adipose tissues. By doing so, the small oil bodies provide ample surface area for the binding of lipase, so that a rapid mobilization of the storage oil can occur in postgermination. The partition of the oils into small droplets may be accomplished by a phospholipid layer without specific proteins. Whether or not oleosins can provide an extra stabilizing effect on the oil bodies remains to be seen. In postgermination, lipase synthesized on free polyribosomes attaches itself specifically to the membrane of the oil bodies, and not the membranes of other subcellular structures. The specific recognition signals should be present on both the lipase and the oil bodies. It is unlikely that the recognition signal on the oil bodies resides on the phospholipids, since their membrane does not have uncommon phospholipids. Thus, it is logical to speculate that the oleosins serve the role of providing a specific binding signal for the lipase in postgermination. If so, the hydrophilic portions of the oleosins exposed to the cytosol are likely candidates for binding sites for the lipase. Further testing of this binding hypothesis is warranted.

Acknowledgement

Our work has been supported by the National Science Foundation, the US Department of Agriculture, and Pioneer Hi-Bred International, Inc.

References

Bergfeld, R., Hong, Y. N., Kühl, T. & Schopfer, P. (1978). Formation of oleosomes (storage lipid bodies) during embryogenesis and their breakdown during seedling development in cotyledons of *Sinapis alba* L. *Planta* **143**, 297–307.

Cao, Y. Z. & Huang, A. H. C. (1986). Diacylglycerol acyltransferase in maturing oil seeds of maize and other species. *Plant Physiol.* **82**, 813–20.

Chou, P. Y. & Fasman, G. D. (1978). Empirical predictions of protein conformation. *Ann. Rev. Biochem.* **47**, 251–76.

Driscoll, D. M. & Getz, G. S. (1986). Molecular and cell biology of lipoprotein biosynthesis. *Meth. Enzymol.* **128**, 41–70.

Engleman, D. M. & Steitz, T. A. (1981). The spontaneous insertion of proteins into and across membranes: The helical hairpin hypothesis. *Cell* **23**, 411–22.

Fernandez, D. E., Qu, R., Huang, A. H. C. & Staehelin, L. A. (1988). Immunogold localization of the L3 protein of maize lipid bodies during germination and seedling growth. *Plant Physiol.* **86**, 270–4.

Gurr, M. I. (1980). The biosynthesis of triacylglycerols. In *The Biochemistry of Plants*, vol. 4, ed. P. K. Stumpf & E. E. Conn, pp. 205–48. New York: Academic Press.

Herman, E. M. (1987). Immunogold-localization and synthesis of an oil-body membrane protein in developing soybean seeds. *Planta* **172**, 336–45.

Huang, A. H. C. (1987). Lipases. In *The Biochemistry of Plants*, vol. 9, ed. P. K. Stumpf & E. E. Conn, pp. 91–119. New York: Academic Press.

Huang, A. H. C., Qu, R., Wang, S. M., Vance, V. B., Cao, Y. Z. & Lin, Y. H. (1987). Synthesis and degradation of lipid bodies in the scutella of maize. In *The Metabolism, Structure, and Function of Plant Lipids*, ed. P. K. Stumpf, J. B. Mudd & W. D. Nes, pp. 239–46. New York: Plenum Press.

Lin, Y. H. & Huang, A. H. C. (1984). Purification and initial characterization of lipase from the scutella of corn seedlings. *Plant Physiol.* **76**, 719–22.

Murphy, D. J., Cummins, I. & Kang, A. S. (1989). Synthesis of the major oil-body membrane protein in developing rapeseed (*Brassica napus*) embryos. *Biochem. J.* **258**, 285–93.

Murphy, D. J. & Au, D. M. Y. (1989). A new class of highly abundant apolipoproteins involved in lipid storage of oilseeds. *Biochem. Soc. Trans.* (In press.)

Ory, R. L. (1969). Acid lipase of castor bean. *Lipids* **4**, 177–85.

Qu, R., Wang, S. M., Lin, Y. H., Vance, V. B. & Huang, A. H. C. (1986). Characteristics and biosynthesis of membrane proteins of lipid bodies in the scutella of maize (*Zea mays* L.). *Biochem. J.* **234**, 57–65.

Stymne, S. & Stobart, A. K. (1987). Triacylglycerol biosynthesis. In *The Biochemistry of Plants*, vol. 9, ed. P. K. Stumpf & E. E. Conn, pp. 175–214. New York: Academic Press.

Schwarzenbach, A. M. (1971). Observations on spherosomal membranes. *Cytobiologie* **4**, 415–17.

Vance, V. B. & Huang, A. H. C. (1987). The major protein from lipid bodies of maize. Characterization and structure based on cDNA cloning. *J. Biol. Chem.* **262**, 11275–9.

Vance, V. B. & Huang, A. H. C. (1988). Expression of lipid body protein gene during maize seed development. *J. Biol. Chem.* **263**, 1476–81.

Wang, S. M., Lin, Y. H. & Huang, A. H. C. (1984). Lipase activities in scutella of maize lines having diverse kernel lipid content. *Plant Physiol.* **76**, 837–9.

Wang, S. M. & Huang, A. H. C. (1987). Biosynthesis of lipase in the scutellum of maize kernel. *J. Biol. Chem.* **262**, 2270–4.

Wanner, G., Formanek, H. & Theimer, R. R. (1981). The ontogeny of lipid bodies in plant cells. *Planta* **151**, 109–23.

Yatsu, L. Y. and Jacks, T. J. (1972). Spherosome membranes. Half unit-membranes. *Plant Physiol.* **49**, 937–43.

J. M. LORD, C. HALPIN, M. J. CONDER AND S. D. IRWIN

Biogenesis of protein bodies and glyoxysomes in *Ricinus communis* seeds

The endosperm cells of maturing *Ricinus communis* (castor beans) seeds are the site of synthesis and deposition of reserve materials which are subsequently utilized during seed germination. These reserves include triglycerides, the major source of carbon during early postgerminative growth, and storage proteins and lectins which, following hydrolysis during germination, serve as a source of amino acids for further protein synthesis. During their biosynthesis, storage proteins and lectins are packaged into and stored within single membrane-enclosed organelles called protein bodies. Storage proteins and the organelles which house them are very abundant in the mature seed, where they account for approximately 90% of the total protein in the endosperm tissue. When the seed germinates, stored triglycerides are metabolized into sucrose which nurtures the growing seedling until it becomes photosynthetically competent. The gluconeogenic enzymes responsible for this conversion are synthesized *de novo* from storage protein-derived amino acids, as are the organelles, the glyoxysomes, into which they are packaged.

Protein bodies and glyoxysomes therefore represent organelles which are superficially similar in structure – both are single membrane delimited and contain a dense protein matrix – but which are formed in the endosperm cell at different development stages. Despite this apparent structural similarity, the mechanisms by which these two types of organelle are assembled differ markedly.

(A) Protein bodies

Structure

Ricinus protein bodies are single membrane bound compartments which contain phytin globoids and a single large protein crystalloid within a soluble protein matrix (Tulley & Beevers, 1976; Youle & Huang, 1976). The organelles are 10–15 μm in diameter. Isolation, subfractionation and electrophoretic analysis has shown that the protein bodies contain three major protein fractions. Over 50% of the organellar protein is 11S globulin, insoluble in water but soluble in salt solution, which constitutes the insoluble crystalloid. The M_r of the 11S globulin is approximately 360,000 and it is a hexameric structure containing six different heterodimers (Gifford & Bewley, 1983). Each heterodimer contains a larger, acidic polypeptide (M_r 30–40,000) linked covalently to a smaller, basic polypeptide (M_r 20–30,000) by a single disulphide bond (Gifford & Bewley, 1983).

The protein body matrix is a mixture of two watersoluble protein fractions, the 7S lectins and the 2S albumins. The lectin fraction consists of two components, ricin and *Ricinus communis* agglutinin (RCA) (Olsnes & Pihl, 1982). Ricin, the first plant lectin to be isolated and characterized over a century ago (Stillmark, 1888), is one of the most potently cytotoxic compounds known. This toxic lectin consists of two distinct N-glycosylated polypeptides joined by a disulphide bond. One polypeptide (the A chain) is a highly specific ribosomal RNA N-glycosidase, which irreversibly inactivates 60S ribosomal subunits thereby causing target cell death (Endo *et al.*, 1987). The second polypeptide (the B chain) is a galactose-specific lectin (Pappenheimer, Olsnes & Harper, 1974). Ricin intoxication of intact cells is initiated when the molecule binds to the cell surface via the B chain sugar-binding sites, which interact with exposed galactosyl residues of surface glycoproteins or glycolipids. Following endocytosis via coated pits and coated vesicles (van Deurs *et al.*, 1985), ricin A chain crosses the membrane of an intracellular compartment, possibly from the trans Golgi cisternae (van Deurs *et al.*, 1986), and enters the target cell cytoplasm where it attacks its ribosomal substrate. Intact ricin has a M_r of around 65,000 (the A chain 32,000, the B chain 34,000). RCA has a M_r of 130,000, and consists of two ricin-like heterodimers held together by non-covalent forces. Once again each heterodimer consists of a toxic A chain (M_r 32,000) disulphide linked to a galactose-binding B chain (M_r 36,000). The corresponding polypeptides of ricin and RCA share extensive primary sequence homology (93% for the A chains, 84% for the B chains) (Roberts *et al.*, 1985).

The 2S albumin fraction was initially isolated and characterized in the 1940s by Spies and his colleagues (Spies & Coulson, 1943). These workers identified a 2S protein fraction as a major allergen present in *Ricinus* seeds, since it was well known that people who handled these seeds regularly often developed a severe allergic response. Subsequently Youle & Huang (1978) showed that the allergen fraction described by Spies and co-workers was indeed the major 2S albumin storage protein. This has been confirmed by the recent demonstration that IgE antibodies in the sera of castor bean-sensitive patients interact strongly and preferentially with the 2S albumin when exposed to total homogenates from mature *Ricinus* seeds (Thorpe *et al.*, 1988). The major component of the 2S albumin fraction has been purified and its primary sequence has been determined (Sharief & Li, 1982). This protein is also a heterodimer consisting of a 7,000 polypeptide joined via a disulphide bond to a 4,000 polypeptide. Both polypeptides are rich in glutamine and cysteine residues and share significant sequence homology with 2S albumin storage proteins from a variety of other plant sources (Allen *et al.*, 1987).

One obvious feature of the major protein components of *Ricinus* protein bodies is that they are comprised of one (ricin, 2S albumin), two (RCA) or six (11S globulin) heterodimers.

Protein body membranes have also been isolated and characterized in terms of their lipid and protein content (Mettler & Beevers, 1979).

Biogenesis

(i) Content proteins

The heterodimeric proteins of the protein body matrix or crystalloid share many common features including the developmental stage at which synthesis occurs, the intracellular site of synthesis, and in transport to and processing within the protein bodies. For this reason a description of the biosynthesis of ricin, the most thoroughly characterized content protein of *Ricinus* protein bodies, will serve as an example for all. The gene encoding ricin has been cloned, showing that both subunits of ricin are synthesized together within a single preproprotein precursor (Lamb, Roberts & Lord, 1985; Halling *et al.*, 1985). Preproricin comprises a 35 amino acid N-terminal extension, which includes a signal sequence, which precedes the A chain sequence (267 amino acids) which is, in turn, separated from the B chain sequence (262 amino acids) by a 12 amino acid linking peptide. Ricin is a member of a small multigene family, which includes at least one pseudogene (Tregear, J., unpublished data), and is most actively expressed during the later stages of seed maturation (Roberts & Lord, 1981a).

In the maturing endosperm cell, translation of preproricin mRNA takes place on membrane-bound polysomes and the nascent polypeptide is co-translationally transported into the rough endoplasmic reticulum (ER) lumen. Segregation is accompanied by cleavage of the N-terminal signal sequence, disulphide bond formation and core glycosylation. Because of heterogeneity of glycosylation, segregated ricin (and RCA) precursors appear as a group of polypeptides of M_r 64–68,000. This has been demonstrated by both *in vitro* (Roberts & Lord, 1981b) and *in vivo* (Lord, 1985a) experiments; synthesis and segregation *in vitro* is illustrated in Fig. 1. From the ER, proricin moves to the Golgi apparatus where post-translational modification of the oligosaccharide side chains, including the addition of fucose, occurs (Lord, 1985b). Proricin is transported from the Golgi apparatus to the protein bodies within a population of transporting vesicles (Lord, 1985a). Having reached the protein bodies, proricin is endoproteolytically cleaved to yield the disulphide linked A and B polypeptides (Harley & Lord, 1985). Delivery into the protein body is effected by the fusion of the transporting vesicle membrane with the protein body membrane and the discharge of the vesicle contents into the protein body matrix. The transporting vesicle contents also include proRCA, proglobulins and proalbumins, all of which are cleaved in the protein body to their mature, heterodimeric forms (Lord, 1985a). The processing endoprotease(s) functions at acid pH, as do the protein body hydrolases that ultimately degrade the storage proteins and lectins during seed germination. Protein bodies can, therefore, be considered similar to other hydrolytic cellular compartments such as animal lysosomes or yeast vacuoles. Proteins destined for these latter two compartments have specific targeting signals; mannose-6-phosphate for lysosomal proteins (Creek & Sly, 1984) and a targeting peptide for yeast vacuolar proteins (Valls et al., 1987). In the case of protein body components, it is not known what parts of the precursors, if any, act as targeting sequences for intracellular transport. However, it is interesting to note that after signal peptide cleavage proricin still has an N-terminal extension which is post-translationally removed in the protein body (Roberts, Lamb & Lord, 1987). Post-translational N-terminal processing has been observed during the biosynthesis of other plant protein body matrix proteins (Crouch et al., 1983; Graham et al., 1985). The synthesis, processing and intracellular transport of ricin is illustrated schematically in Fig. 2.

Our recent studies have indicated that the biosynthesis of the *Ricinus* 2S albumin has unusual and interesting features. Rabbit antibodies against the glutamine-rich 2S albumin, a heterodimer of disulphide-linked 4,000 and 7,000 polypeptides, immunoprecipitated a precursor product of apparent M_r 34,000 from the total products formed when *Ricinus* endosperm mRNA was translated *in vitro* (Fig. 1). We have generated full length cDNA clones encoding the 2S albumin precursor. The preproalbumin precursor contains 258 amino acid residues and includes an N-terminal signal sequence (21 residues), and the large (65 residues) and small (34 residue) 2S albumin subunits (Fig. 3). In other words, even after removal of the signal sequence, the 2S albumin subunits (99 residues) represent only approximately 40% of the segregated proricin precursor (237 residues). Clearly this suggests that either (a), proces-

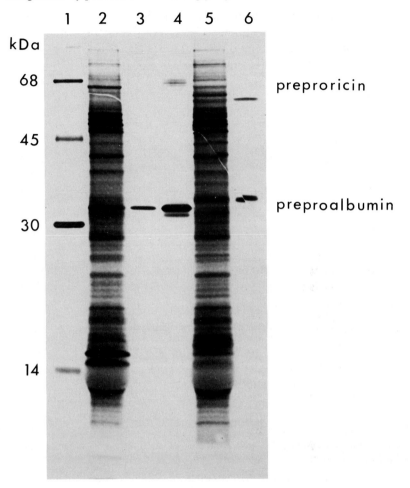

Fig. 1. *In vitro* synthesis of preproricin and preproalbumin. Ripening castor bean mRNA was translated in rabbit reticulocyte lysates in the presence or absence of dog pancreatic microsomes. Precursors were recovered by immunoprecipitation. Lane 1, molecular weight markers; lanes 2 and 5, total products synthesized in the presence or absence of microsomes; lanes 3 and 4, after synthesis in the presence of microsomes, the microsomes were recovered by centrifugation and precursors located in the supernatant (lane 3) or microsome pellet (lane 4). Lane 6, immunoprecipitate from total products synthesized in the absence of microsomes.

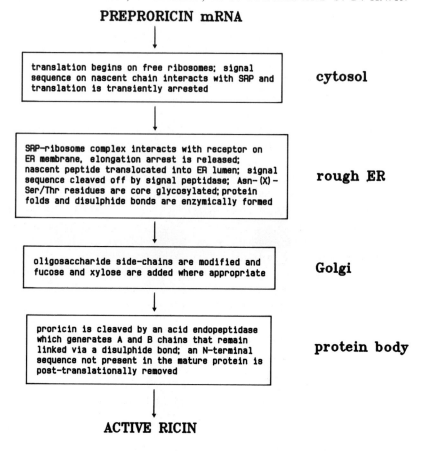

PREPRORICIN mRNA

translation begins on free ribosomes; signal
sequence on nascent chain interacts with SRP and
translation is transiently arrested

cytosol

SRP—ribosome complex interacts with receptor on
ER membrane, elongation arrest is released;
nascent peptide translocated into ER lumen; signal
sequence cleaved off by signal peptidase; Asn-(X)-
Ser/Thr residues are core glycosylated; protein
folds and disulphide bonds are enzymically formed

rough ER

oligosaccharide side-chains are modified and
fucose and xylose are added where appropriate

Golgi

proricin is cleaved by an acid endopeptidase
which generates A and B chains that remain
linked via a disulphide bond; an N-terminal
sequence not present in the mature protein is
post-translationally removed

protein body

ACTIVE RICIN

Fig. 2. The biosynthesis, intracellular transport and processing of ricin.

signal
sequence

small   large

2S albumin subunits

Fig. 3. Schematic representation of prepro 2S albumin.

sing is very inefficient in terms of the retention of newly synthesized
protein, or (b), since the precursor sequence, potentially lost during
processing, is itself rich in glutamine and cysteine residues, the possibility
exists that a second storage protein, in addition to the characterized 2S
albumin, is derived from a common precursor.

(ii) Membrane protein

The demonstration that protein body content proteins follow the ER→Golgi→protein body route suggest that integral proteins of the protein body membrane could be synthesized and transported in exactly the same way. Protein body membrane proteins co-translationally inserted into the rough ER membrane could pass through the same pathway as content proteins, being present finally in the membrane of the dense transporting vesicles that fuse with the protein body membrane. This contention is supported by experimental evidence. Protein body membrane proteins (in particular a predominant 25 kDa polypeptide) synthesized *in vitro* and identified by immunoprecipitation with anti-protein body membrane antibodies, are efficiently co-translationally inserted into rough ER membranes. These proteins do not appear to be synthesized as precursors with cleavable N-terminal signal sequences suggesting they may possess internal signal sequences.

In summary, the ER appears to have an important role in the biogenesis of protein bodies as the site of synthesis and segregation of protein body content proteins and the site of synthesis and membrane insertion in the case of protein body membrane proteins.

(B) Glyoxysomes

Structure

Glyoxysomes are small organelles, $0 \cdot 5 – 1 \cdot 0\,\mu m$ in diameter, that are bounded by a single membrane, which encloses an amorphous or granular matrix (Beevers, 1979). The matrix may also contain electron-dense, often semi-crystalline, material known as a core or nucleoid. Glyoxysomes are a specialized form of peroxisomes (microbodies), organelles considered to be ubiquitously present in eukaryotic cells (Kindl & Lazarow, 1982). Almost without exception peroxisomes can be biochemically characterized by the possession of one or more hydrogen peroxide-generating oxidases and hydrogen peroxide-degrading catalase. In addition an inducible fatty acid β-oxidation system is present in the peroxisomes of all organisms analysed to date, although the activity of this system varies between tissues.

Apart from morphological and biochemical similarities, peroxisomes from different organisms and tissues can contain very different complements of enzymes, reflecting the diverse metabolic functions that these organelles can perform. In germinating fat-storing seeds these organelles play an essential role in gluconeogenesis from fats (Beevers, 1979). The glyoxylate cycle is the key to this metabolic activity and these specialized peroxisomes are therefore known as glyoxysomes.

Glyoxysomes and their enzymic components are either absent or present at very low levels in mature fat-storing seeds. During early post-germinative growth there is a rapid and large scale *de novo* synthesis of glyoxysomes. In the case of *Ricinus* seeds growing at 30°C, the glyoxysomal content of the endosperm cells reaches a maximal level after 4–5 days before rapidly declining (Gerhardt & Beevers, 1970). The conversion of stored fat to sucrose is the dominant metabolic event at this stage of development, providing the developing shoots with a source of carbon and energy until the first green leaves appear and photosynthesis begins. The glyoxysomes, which at the peak of their activity represent approximately 20% of the total particulate cellular protein, contain the enzymes of the β-oxidation pathway and those of the glyoxylate cycle. This compartmentation scheme ensures that acetyl-CoA generated by the β-oxidation of fatty acids avoids the oxidative decarboxylations of the mitochondrial citric acid cycle and is instead converted exclusively to succinate via the glyoxylate cycle. As a result up to 75% of the fatty acid carbon can be ultimately recovered in the cytoplasm as sucrose (Beevers, 1979).

Early evidence indicated that certain glyoxysomal content proteins, including malate synthase and citrate synthase (Huang & Beevers, 1973), were peripherally associated with the organellar membrane. This conclusion emerged from the demonstration that when isolated glyoxysomes were deliberately broken by osmotic shock and most of the soluble matrix enzymes were largely released, the bulk of certain enzymes such as malate synthase was recovered in the membrane fraction, requiring washes in buffers containing KCl for effective solubilization (Huang & Beevers, 1973; Huang, Trelease & Moore, 1983). More recently, enzyme cytochemical (Trelease, 1987) and immunochemical (Titus & Becker, 1985) studies have indicated that malate synthase is distributed throughout the glyoxysomal matrix, and is not specifically associated with the membrane (Chapman, Turley & Trelease, 1989).

Ricinus glyoxysomal membranes have been purified and shown to contain enzymes and cytochromes (Donaldson *et al.*, 1981) including, as a major integral component, an alkaline lipase (Muto & Beevers, 1974; Maeshima & Beevers, 1985).

Biogenesis

(i) Content proteins
Early experimental evidence was interpreted as support for an ER-vesiculation model (Beevers, 1979) in which glyoxysomal membrane and matrix proteins were synthesized on membrane-bound polysomes. These

proteins were believed to be inserted into or translocated across the rough ER membrane by a mechanism similar to that described above for the synthesis and segregation of protein body polypeptides. One highly suggestive result supporting this model was the demonstration that a significant proportion of castor bean endosperm malate synthase was associated with the ER membrane fraction (Gonzalez & Beevers, 1976), where it was believed to have been segregated into the microsomal vesicles (Lord & Bowden, 1978). More recent re-examination of this earlier work has shown that malate synthase was fortuitously associated with the ER because enzyme aggregates co-sedimented with the ER vesicles on sucrose gradients (Kindl, 1982; Chapman *et al.*, 1989). Indeed numerous attempts to demonstrate that glyoxysomal matrix proteins are synthesized on membrane bound polysomes and reach the glyoxysomes by vesiculation from the ER have consistently failed (reviewed by Trelease, 1984). It is now firmly accepted that glyoxysomal matrix proteins, including malate synthase, in common with content proteins of all types of microbodies (peroxisomes, glycosomes), are synthesized on free polysomes (Borst, 1986). Classical cellular fractionation techniques have been used to separate free cytosolic polysomes from membrane bound polysomes. Translation of these two fractions (or the mRNA derived from them) *in vitro* has, in all cases examined to date, shown mRNAs encoding microbody matrix proteins to be predominantly associated with free polysomes. One result obtained in the case of *Ricinus* glyoxysomal isocitrate lyase is illustrated in Fig. 4. In addition, radiolabelling studies followed by tissue fractionation have shown that *in vivo* microbody matrix proteins can be initially detected in the soluble cytosolic fraction before they accumulate in the microbody fraction. These data show that microbody matrix proteins are synthesized on free polysomes, released into the cytosol from where they are post-translationally imported into the microbodies (Borst, 1986). The ER appears to play no role in this process (Lazarow & Fujiki, 1985). Direct evidence for post-translational import has been obtained *in vitro* for both glyoxysomal (Zimmerman & Neupert, 1980; Kindl, 1982) and peroxisomal (Lazarow & Fujiki, 1985) matrix proteins.

While the co-translational segregation of protein body content proteins is mediated by a cleavable N-terminal signal sequence, most microbody matrix proteins are not made as longer precursors. Rather, most microbody matrix proteins are synthesized at their mature molecular weights. In the case of mammalian peroxisomal catalase, the N-terminal sequences of newly synthesized protein and mature protein have been determined and found to be identical (Lazarow & Fujiki, 1985). Microbody proteins are, therefore, different to nuclear-encoded

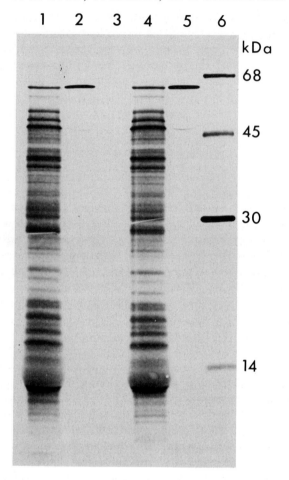

Fig. 4. *In vitro* synthesis of isocitrate lyase. Ripening castor bean mRNA was translated in rabbit reticulocyte lysates in the presence or absence of dog pancreatic microsomes. Isocitrate lyase was recovered by immunoprecipitation. Lanes 1 and 4, total products synthesized in the presence or absence of microsomes, respectively; lanes 2 and 3, after synthesis in the presence of microsomes, the microsomes were recovered by centrifugation and isocitrate lyase was located in the supernatant (lane 2) or microsomal pellet (lane 3); lane 5, immunoprecipitate from total products synthesized in the absence of microsomes; lane 6, molecular weight markers. (From Roberts & Lord, 1981c. Used with permission from European Journal of Biochemistry.)

mitochondrial and chloroplast proteins, which are usually initially synthesized as larger molecular weight precursors with N-terminal extension or targeting sequences which direct the proteins to the correct organelle (Verner & Schatz, 1988). These targeting sequences are cleaved during or after organelle import.

Although synthesis at the mature molecular weight appears to be the norm for most microbody matrix proteins, there are exceptions. One such exception is glyoxysomal malate dehydrogenase which is synthesized as a precursor about 8 kDa larger than the mature protein (Gietl & Hoch, 1982). The function of the extra sequences present in microbody precursors is unclear, as there is no convincing evidence that they are responsible for targeting or import, although this has often been suggested by analogy with the import mechanisms of chloroplasts and mitochondria.

For those microbody matrix proteins not synthesized as precursors, the targeting signals that direct these proteins to the microbodies must be present within the mature protein. Recently recombinant DNA techniques have been successfully used to identify such targeting sequences. Firefly luciferase, which catalyses a light-producing bioluminescent reaction, is located in peroxisomes in cells of the firefly lantern organ. When cDNA encoding luciferase, a gene for which there is no mammalian homologue, is introduced into and expressed in mammalian cells, the enzyme is found in the peroxisomes of these cells (Keller, Gould & Subramani, 1987). By deleting the luciferase gene it has been possible to identify regions of the luciferase protein that are necessary and sufficient for its transport into peroxisomes. The signal is a short amino acid sequence at the extreme carboxy terminus of luciferase (Gould, Keller & Subramani, 1987). Moreover gene fusion experiments which resulted in the luciferase peroxisomal targeting sequence being fused to the carboxy-terminus of mouse dehydrofolate reductase or bacterial chloramphenicol acetyltransferase also directed these passenger proteins to peroxisomes. Subsequently targeting signals for several peroxisomal matrix proteins have been shown to reside at the extreme carboxy terminus (Gould, Keller & Subramani, 1988; Miyazawa *et al.*, 1989). A comparison of the carboxy termini of several microbody enzymes (Fig. 5) shows that a tripeptide of the sequence Ser-Lys/His-Leu is frequently, but not invariably, present in these regions, often at the extreme carboxy terminus. While this type of experimental approach has not yet been applied to plant glyoxysomal matrix proteins, it is reasonable to assume that similar targeting sequences will exist, particularly as the primary sequence for cucumber glyoxysomal malate synthase ends in Ser-Lys-Leu (Smith & Leaver, 1986).

```
Rat acylCoA ox.      -Y H K H L K P L Q│S K L│
Firefly luciferase   -L I K A K K G G K│S K L│
Rat bifunctional     -W Q S L A G P H G│S K L│
Soybean uricase      -I Q A S L S R L W│S K L│
Rat uricase          -T G T V R R K L P│S K L│
Malate synthase      -I V I H H P R E L│S K L│

Pig amino acid ox.   -E R N L L T M P P S H L
Human catalase       -G│S H L│A A R E K A N L

Rat thiolase         -M G A A A V F E Y P Q N
Isocitrate lyase     -S A G S E V V A K A R M
```

Fig. 5. Carboxy terminal sequences of several microbody proteins.

(ii) Membrane proteins

The biosynthesis of two microbody integral membrane proteins has been examined to date. These are a major 22 kDa polypeptide from rat liver peroxisomes (Fujiki, Rachubinski & Lazarow, 1984) and the alkaline lipase from *Ricinus* glyoxysomes (Halpin, Conder & Lord, unpublished). Both of these integral membrane proteins were found to be synthesized on free polysomes and must therefore enter the organelle membrane post-translationally. There is no evidence that either of these membrane proteins is synthesized as a higher molecular weight precursor. This is important because microbody membranes were originally proposed to arise directly from the ER membrane by vesiculation (Goldman & Blobel, 1978).

If all glyoxysomal proteins are incorporated post-translationally, what can act as receptor(s) for their recognition? An answer to this question might be provided by reports that glycoproteins are present in the glyoxysomal membrane (Bergner & Tanner, 1981; Lord & Roberts, 1983). Because N-glycosylation is a co-translational event which occurs exclusively in the rough ER, these results suggest that at least some glyoxysomal proteins may reach the organelle by membrane flow from the ER. These co-translationally incorporated glycoproteins might then act as receptors facilitating the post-translational uptake of other integral membrane proteins and matrix proteins. However glycoproteins, if present in the glyoxysomal membrane, represent only very minor constituents and do not coincide with any of the major stained bands or with newly synthesized [35S] methionine-labelled polypeptides on SDS-polyacrylamide gels (Bergner & Tanner, 1981). In addition, glyoxysome preparations isolated from sucrose gradients can be contaminated with small amounts of other cellular membranes, including ER vesicles. Reports of small amounts of glycoproteins in glyoxysomal membranes must, therefore, be viewed cautiously. The major glyoxysomal integral membrane proteins appear to be incorporated into the organelle post-translationally without any ER involvement.

Conclusions

Although *Ricinus* protein bodies and glyoxysomes are superficially similar in structure and are assembled in the same cells at different developmental stages, they are assembled by different mechanisms. Protein body matrix and membrane proteins are synthesized on membrane bound polysomes and pass through the ER and Golgi apparatus on route to the protein bodies. Glyoxysomal matrix and membrane proteins are synthesized on free polysomes, released into the cytosol from where they are post-translationally imported into glyoxysomes. Details of the post-translational import signals are beginning to emerge.

References

Allen, R. D., Cohen, E. A., Vonder Haar, R. A., Adams, C. A., Ma, D. P., Nessler, C. L. & Thomas, T. L. (1987). Sequence and expression of a gene encoding an albumin storage protein in sunflower. *Molec. Gen. Genet.* **210**, 211–18.

Beevers, H. (1979). Microbodies in higher plants. *Ann. Rev. Plant Physiol.* **30**, 159–97.

Bergner, U. & Tanner, W. (1981). Occurrence of several glycoproteins in glyoxysomal membranes of castor beans. *FEBS Lett.* **131**, 68–72.

Borst, P. (1986). How proteins get into microbodies (peroxisomes, glyoxysomes, glycosomes). *Biochim. Biophys. Acta* **866**, 179–204.

Chapman, K. D., Turley, R. B. & Trelease, R. N. (1989). Relationship between cotton seed malate synthase aggregation behaviour and suborganelle location in glyoxysomes and endoplasmic reticulum. *Plant Physiol.* **89**, 352–9.

Creek, K. E. & Sly, W. S. (1984). The role of the phosphomannosyl receptor in the transport of acid hydrolases to lysosomes. In *Lysosomes in Pathology and Biology* ed. J. T. Dingle, R. T. Dean & W. S. Sly, pp. 63–82. New York: Elsevier/North Holland.

Crouch, M. L., Tenbarge, K. M., Simon, A. E. & Ferl, R. (1983). cDNA clones for *Brassica napus* seed storage proteins. *J. Molec. Appl. Genet.* **2**, 272–83.

Donaldson, R. P., Tulley, R. E., Young, A. O. & Beevers, H. (1981). Organelle membranes from germinating castor bean endosperm. II

Enzymes, cytochromes and permeability of the glyoxysomal membrane. *Plant Physiol.* **67**, 21–5.

Endo, Y., Mitsui, K., Motizuki, M. & Tsurugi, K. (1987). The mechanism of action of ricin and related toxins on eukaryotic ribosomes. *J. Biol. Chem.* **262**, 5908–12.

Fujiki, Y., Rachubinski, R. A. & Lazarow, P. B. (1984). Synthesis of a major integral membrane polypeptide of rat liver peroxisomes on free polysomes. *Proc. Natn. Acad. Sci. USA* **81**, 7127–31.

Gerhardt, B. & Beevers, H. (1970). Developmental studies on glyoxysomes from castor bean endosperm. *J. Cell Biol.* **40**, 94–102.

Gietl, C. & Hoch, B. (1982). Organelle-bound malate dehydrogenase isoenzymes are synthesized as higher molecular weight precursors. *Plant Physiol.* **70**, 483–7.

Gifford, D. J. & Bewley, J. D. (1983). An analysis of the subunit structure of the crystalloid protein complex from castor bean endosperm. *Plant Physiol.* **72**, 376–81.

Goldman, B. M. & Blobel, G. (1978). Biogenesis of peroxisomes: intracellular site of synthesis of catalase and uricase. *Proc. Natn. Acad. Sci. USA* **75**, 5066–70.

Gonzalez, E. & Beevers, H. (1976). Role of the endoplasmic reticulum in glyoxysome formation in castor bean endosperm. *Plant Physiol.* **57**, 406–9.

Gould, S. J., Keller, G. A. & Subramani, S. (1987). Identification of a peroxisomal targeting signal at the carboxyl terminus of firefly luciferase. *J. Cell Biol.* **105**, 2923–31.

Gould, S. J., Keller, G. A. & Subramani, S. (1988). Identification of peroxisomal targeting signals located at the carboxy terminus of four peroxisomal proteins. *J. Cell Biol.* **107**, 897–905.

Graham, J. S., Pearce, G., Merryweather, J., Titani, K., Ericsson, L. H. & Ryan, C. A. (1985). Wound-induced proteinase inhibitors from tomato leaves. II The cDNA-deduced primary structure of pre-inhibitor II. *J. Biol. Chem.* **260**, 6561–4.

Halling, K. C., Halling, A. C., Murray, E. E., Ladin, B. F., Houston, L. L. & Weaver, R. F. (1985). Genomic cloning and characterization of a ricin gene from *Ricinus communis*. *Nucleic Acids Res.* **13**, 8019–34.

Harley, S. M. & Lord, J. M. (1985). *In vitro* endoproteolytic cleavage of castor bean lectin precursors. *Plant Sci.* **41**, 111–16.

Huang, A. H. C. & Beevers, H. (1973). Localization of enzymes within microbodies. *J. Cell Biol.* **58**, 379–89.

Huang, A. H. C., Trelease, R. N. & Moore, T. S. (1983). *Plant Peroxisomes* New York: Academic Press.

Keller, G. A., Gould, S. J. & Subramani, S. (1987). Firefly luciferase is targeted to peroxisomes in mammalian cells. *Proc. Natn. Acad. Sci. USA* **84**, 3264–8.

Kindl, H. (1982). Glyoxysome biogenesis via cytosolic pools in cucumber. *Ann. N. Y. Acad. Sci.* **386**, 314–28.

Kindl, K. & Lazarow, P. B. (1982). Peroxisomes and glyoxysomes. *Ann. N. Y. Acad. Sci.* **386**, 1–548.

Lamb, F. I., Roberts, L. M. & Lord, J. M. (1985). Nucleotide sequence of cloned cDNA coding for preproricin. *Eur. J. Biochem.* **185**, 265–70.

Lazarow, P. B. & Fujiki, Y. (1985). Biogenesis of peroxisomes. *Ann. Rev. Cell Biol.* **1**, 489–530.

Lord, J. M. (1985a). Synthesis and intracellular transport of lectin and storage protein precursors from castor bean. *Eur. J. Biochem.* **146**, 403–9.

Lord, J. M. (1985b). Precursors of ricin and *Ricinus communis* agglutinin. Glycosylation and processing during synthesis and intracellular transport. *Eur. J. Biochem.* **146**, 411–16.

Lord, J. M. & Bowden, L. (1978). Evidence that glyoxysomal malate synthase is segregated by the endoplasmic reticulum. *Plant Physiol.* **61**, 266–70.

Lord, J. M. & Roberts, L. M. (1983). Formation of glyoxysomes. *Int. Rev. Cytol.* **15**, 115–56.

Maeshima, M. & Beevers, H. (1985). Purification and properties of glyoxysomal lipase from castor bean. *Plant Physiol.* **79**, 489–93.

Mettler, I. J. & Beevers, H. (1979). Isolation and characterization of the protein body membrane of castor beans. *Plant Physiol.* **64**, 506–11.

Miyazawa, S., Osumi, T., Hushimoto, T., Ohno, K., Miura, S. & Fujiki, Y. (1989). Peroxisome targeting signal of rat liver acyl coenzyme A oxidase resides at the carboxy terminus. *Molec. Cell. Biol.* **9**, 83–91.

Muto, S. H. & Beevers, H. (1974). Lipase activities in castor bean endosperm during germination. *Plant Physiol.* **54**, 23–28.

Olsnes, S. & Pihl, A. (1982). Toxic lectins and related proteins. In *Molecular Action of Toxins and Viruses*, ed. P. Cohen & S. van Heyningen, pp. 51–105. New York: Elsevier.

Pappenheimer, A. M., Olsnes, S. & Harper, A. A. (1974). Lectins from *Abrus precatorius* and *Ricinus communis*. *J. Immunol.* **113**, 835–41.

Roberts, L. M., Lamb, F. I. & Lord, J. M. (1987). Biosynthesis and molecular cloning of ricin and *Ricinus communis* agglutinin. In *Membrane-mediated Cytotoxicity*, ed. B. Bonavida & R. J. Collier, pp. 73–82. New York: Alan R. Liss.

Roberts, L. M., Lamb, F. I., Pappin, D. J. C. & Lord, J. M. (1985). The primary sequence of *Ricinus communis* agglutinin. Comparison with ricin. *J. Biol. Chem.* **260**, 15682–6.

Roberts, L. M. & Lord, J. M. (1981a). Protein biosynthetic capacity in the endosperm tissue of ripening castor bean seeds. *Planta* **152**, 420–7.

Roberts, L. M. & Lord, J. M. (1981b). The synthesis of *Ricinus com-*

munis agglutinin. Co-translational and post-translational modification of agglutinin polypeptides. *Eur. J. Biochem.* **119**, 31–41.

Roberts, L. M. & Lord, J. M. (1981c). Synthesis and post-translational segregation of glyoxysomal isocitrate lyase from castor bean endosperm. *Eur. J. Biochem.* **119**, 43–9.

Sharief, F. S. & Li, S. S-L. (1982). Amino acid sequence of small and large subunits of seed storage protein from *Ricinus communis. J. Biol. Chem.* **257**, 14753–9.

Smith, S. M. & Leaver, C. J. (1986). Glyoxysomal malate synthase of cucumber: molecular cloning of a cDNA and regulation of enzyme synthesis during germination. *Plant Physiol.* **81**, 762–7.

Spies, J. R. & Coulson, E. J. (1943). The chemistry of allergens: isolation and purification of active protein-polysaccharide fraction CB-1A from castor beans. *J. Am. Chem. Soc.* **65**, 1720–6.

Stillmark, H. (1988). Uber Ricin, eines gifiges Ferment ans den Samen von *Ricinus communis* L. und andersen Euphorbiaceen. Inaugural Dissertation, University of Dorpat, Estonia.

Thorpe, S. C., Kemeny, M. D., Panzani, R. C., McGurl, B. & Lord, J. M. (1988). Allergy to castor bean II. Identification of the major allergens in castor bean seeds. *J. Aller. Clin. Immunol.* **82**, 67–72.

Titus, D. E. & Becker, W. M. (1985). Investigation of the glyoxysome-peroxisome transition in germinating cucumber cotyledons using double-label immunoelectron microscopy. *J. Cell Biol.* **101**, 1288–99.

Trelease, R. N. (1984). Biogenesis of glyoxysomes. *Ann. Rev. Plant Physiol.* **35**, 321–47.

Trelease, R. N. (1987). Malate synthase. In *Handbook of Plant Cytochemistry*, ed. K. Vaughn, pp. 133–47. Boca Rato: CRC Press.

Tulley, R. E. & Beevers, H. (1976). Protein bodies of castor bean endosperm. Isolation, fractionation and characterization of protein components. *Plant Physiol.* **58**, 710–16.

Valls, L. A., Hunter, C. P., Rothman, J. H. & Stevens, T. H. (1987). Protein sorting in yeast, the localization determinant of yeast vacuolar carboxypeptidase Y resides in the propeptide. *Cell* **48**, 887–97.

van Deurs, B., Ryde-Petersen, L., Sundan, A., Olsnes, S. & Sandvig, K. (1985). Receptor-mediated endocytosis of a ricin-colloidal gold conjugate in Vero cells. *Expl. Cell Res.* **159**, 287–305.

van Deurs, B., Tonnessen, T. I., Petersen, D. W., Sandvig, K. & Olsnes, S. (1986). Routing of internalized ricin and ricin conjugates to the Golgi complex. *J. Cell. Biol.* **102**, 37–45.

Verner, K. & Schatz, G. (1988). Protein translocation across membranes. *Science* **241**, 1307–13.

Youle, R. J. & Huang, A. H. C. (1976). Protein bodies from the endosperm of castor bean. Subfractionation, protein components, lectins and changes during germination. *Plant Physiol.* **58**, 703–9.

Youle, R. J. & Huang, A. H. C. (1978). Evidence that the castor bean allergens are the albumin storage proteins in the protein bodies of castor bean. *Plant Physiol.* **61**, 1040–42.

Zimmerman, R. & Neupert, W. (1980). Biogenesis of glyoxysomes. Synthesis and intracellular transfer of isocitrate lyase. *Eur. J. Biochem.* **112**, 225–33.

D. T. DENNIS, S. BLAKELEY AND
S. CARLISLE

Isozymes and compartmentation in leucoplasts

Introduction

Until recently it had been assumed that the metabolism of plants would have many properties in common with, if not identical to, those found in bacteria and animals. It has now been shown that the principal biochemical pathways have many similar components but as more information is obtained it has become increasingly obvious that there are fundamental differences. A basic feature of plants is that they are highly compartmented and have plastids, organelles unique to this kingdom.

Plastids are present in all the tissues of a plant but vary greatly in structure and function. The most studied plastids are the chloroplasts that are found principally in leaves but can also be identified in storage tissues, such as seeds. Even these organelles exhibit specialized activities which depend upon factors such as the primary metabolic activity of the organ or its age. In non-photosynthetic tissues such as roots and storage organs the plastids can have very different morphologies and biochemical activities.

Properties of plastids

Although plastids have very different properties in various tissues it appears that all plastids from an individual plant have the same genome and develop from the same precursor proplastids (Dennis et al., 1985). The coding capacity of the plastid genome is for about 130 proteins, many of which have still to be identified. Since the complement of enzymes within the plastid is much higher than could be encoded by a genome of this size, most of the enzymes in the plastid must be encoded in the nucleus, translated in the cytosol and imported into the organelle. This process has been studied in detail in the case of the small subunit of Rubisco (Highfield & Ellis, 1978). The uptake of proteins into the non-photosynthetic plastids that are found in the endosperm of the castor oil plant appears to proceed by a similar if not identical mechanism to that found in chloroplasts (Boyle, Hemmingsen & Dennis, 1986), suggesting that this mechanism is universal, although this will not be confirmed until more proteins have been studied. In addition to the imported enzyme complement, mRNA levels from genes that are encoded in the plastid genome may vary in different plastids (Dennis, 1989). This is likely to be a result of differences in the stability of the messenger rather than differences in transcription (Deng & Gruissem, 1987).

The anabolic activity of a tissue will be determined to a large extent by the metabolic capacity of the plastids within that tissue since much of the biosynthetic activity of a plant occurs in these organelles. However, some enzyme activities that are found in the plastid may also be duplicated in other parts of the cell, especially the cytosol (Dennis & Miernyk, 1982). When the same enzyme activity is found in different cell compartments it is usually catalyzed by isozymes that are specific for a given compartment, although in most cases this has not been proven by genetic or molecular analysis (Weeden & Gottlieb, 1980).

Leucoplasts in the developing endosperm of the castor plant

A significant contribution to our understanding of isozymes and cellular compartmentation has been made through the study of the endosperm of the developing castor seed (Dennis & Miernyk, 1982; Dennis, 1989). This tissue stores large amounts of oil for use by the seedling during germination. It has been shown that the site of synthesis of the fatty acids for the storage oil is a colourless plastid that has been termed a leucoplast (Zilkey & Canvin, 1972). It has now been demonstrated that fatty acid biosynthesis is confined to plastids in all plant tissues (Ohlrogge, Kuhn & Stumpf, 1979). The carbon for this biosynthesis is supplied by the action of a pyruvate dehydrogenase complex that is located in the plastid and which converts pyruvate to acetyl-CoA, the immediate precursor of the fatty acid synthetase complex (Reid *et al.*, 1977). That this complex is active *in vivo* has been shown by the observation that isolated leucoplasts, when incubated with radioactive pyruvate, will incorporate this label into fatty acids (Miernyk & Dennis, 1983).

Pyruvate could be supplied to the leucoplast by a variety of routes and it is likely that several of them are used in different plants and in the various tissues of the same plant. *In vitro*, pyruvate can be taken up by plastids; in the cell this pyruvate can be formed by cytosolic glycolysis. However, it has been demonstrated that plastids from a variety of sources contain at least some of the enzymes of the glycolytic pathway and could be capable of generating the pyruvate internally (Dennis & Miernyk, 1982). The leucoplasts from the developing castor endosperm have been shown to contain all the enzymes of the glycolytic pathway (Simcox *et al.*, 1977; Dennis & Miernyk, 1982) and isolated leucoplasts can incorporate 3-phosphoglyceric acid and hexose into fatty acids suggesting that the pathway can function in supplying carbon to the pyruvate dehydrogenase complex (Miernyk & Dennis, 1983). A diagram of this pathway is presented in Fig. 1.

The glycolytic pathway, in addition to supplying carbon for fatty acid biosynthesis, will also generate reducing power and ATP. Fatty acid biosynthesis by isolated leucoplasts requires externally added ATP (Miernyk & Dennis, 1983). This requirement can be replaced by phosphoenol pyruvate (PEP) which generates ATP within the organelle via the operation of leucoplast pyruvate kinase (Ireland, DeLuca & Dennis, 1980; Boyle, Hemmingsen & Dennis, 1990). In fact, the rate of fatty acid biosynthesis is much higher with internally generated ATP than when it is added externally (Boyle *et al.* 1990). 2-phosphoglycerate will also generate ATP for fatty acid biosynthesis although less effectively than PEP (Boyle *et al.*, 1990). Similar results to these have been obtained with chromoplasts (Kleinig & Liedvogel, 1980).

These data, along with the incorporation results, indicate that hexose, 3-phosphoglycerate, 2-phosphoglycerate, PEP and pyruvate can all cross the leucoplast membrane. Since the leucoplast membrane appears to contain the phosphate translocator (Miernyk & Dennis, 1983), it is reasonable to assume that triose phosphate can also enter the leucoplast. The rate of transport of these various glycolytic intermediates is not known and there is no information to indicate whether any intermediate is used preferentially *in vivo*. The extent of the glycolytic pathway contained within plastids can be very different and this probably depends upon metabolic requirements of the tissues and upon which intermediates are taken up by the plastids.

Phosphoglyceromutase has been reported to be absent in mature chloroplasts (Stitt & ap Rees, 1979) and the level of enolase varies from 30% of the total cellular activity in leucoplasts to zero in mature chloroplasts (Miernyk & Dennis, unpublished). Plastids from other tissues have intermediate amounts of this enzyme. A complete glycolytic pathway has been found in plastids from other plants (e.g. Journet & Douce, 1985) and the level of the enzymes of the pathway seems to correlate with biosynthetic activity.

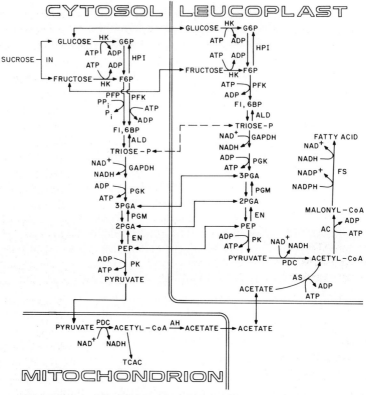

Fig. 1. A diagram of the flow of carbon from sucrose to fatty acid in the developing endosperm of the castor plant. The uptake into the leucoplast of glucose, fructose, 3PGA, pyruvate and acetate has been demonstrated by the incorporation of these compounds into fatty acids by isolated leucoplasts (Miernyk and Dennis, 1983). The uptake of 2-PGA and PEP is inferred by the fact that these compounds can be used to generate the internal ATP required for the incorporation of acetate into fatty acids (Boyle, Hemmingsen & Dennis, 1990). The uptake of triose phosphate has not been demonstrated, but it is assumed to occur on the phosphate transporter. The preferred *in vivo* route of uptake is not known. Only the enzyme cofactors involved in energy and reducing power are indicated. Enzyme abbreviations are as follows: IN, invertase; HK, hexokinase; HPI, hexose phosphate isomerase; PFK, ATP-dependent phosphofructokinase; PFP, PP_i-dependent phosphofructokinase; ALD, aldolase; GAPDH, glyceraldehyde 3-phosphate dehydrogenase; PGK, phosphoglycerate kinase; PGM, phosphoglyceromutase; EN, enolase; PK, pyruvate kinase; PDC, pyruvate dehydrogenase complex; AC, acetyl-CoA carboxylase; AS, acetyl-CoA synthase; FS, fatty acid synthetase; AH, acetyl-CoA hydrolase; TCAC, tricarboxylic acid cycle. Intermediates in the pathways are abbreviated as follows: G6P, glucose 6-phos-

Isozymes in leucoplasts

The glycolytic enzymes that are located in the leucoplast of the developing endosperm can be separated from their cytosolic counterparts, suggesting that they may be isozymes (Ireland & Dennis, 1980; Miernyk & Dennis, 1982). In some cases, such as phosphofructokinase and pyruvate kinase, the leucoplast and cytosolic forms of the enzyme have quite different kinetic and physical properties (Garland & Dennis, 1980; Ireland *et al.*, 1980) whereas in other cases, such as enolase and phosphoglyceromutase, the properties of the isoforms are very similar (Miernyk & Dennis, 1984; Botha & Dennis, 1986). It is not known for the endosperm tissue whether the plastid and cytosolic forms of the glycolytic enzymes are encoded by separate genes and are, therefore, true isozymes. Separate genes for the chloroplast and cytosolic forms of some of these enzymes have been described in other plants (Weeden & Gottlieb, 1980). In animals, it has been shown that the initial transcript of a gene that contains introns can be differentially spliced to give rise, on translation, to more than one enzyme form (Noguchi, Inoue & Tanaka, 1986). Whether such differential splicing of mRNA, or post-translational modification of proteins, plays a role in the differentiation of plastid and cytsolic enzymes is not known.

The present state of our understanding of some plastid and cytosolic isozymes of the glycolytic pathway in developing castor endosperm will be described in the rest of this chapter. Particular reference will be given to recent data available on this system.

Phosphofructokinases

There are two fructose 6-phosphate 1-phosphotransferase enzymes in plants that convert fructose 6-phosphate (Fru-6-P) to fructose 1,6-bisphosphate (Fru-1,6-P$_2$). The first to be described phosphofructokinase (PFK) uses ATP as the phosphoryl donor to Fru-6-P and generates ADP as a product. This enzyme occurs universally in plants, animals and microorgansims. More recently, a second enzyme pyrophosphate dependent fructose-6-phosphate 1-phosphotransferase (PFP) has been described that uses pyrophosphate as the phosphoryl donor and is distributed widely in plants but outside the plant kingdom appears to be found in only a few isolated microorganisms (Carnal & Black, 1983). PFP

Caption for Fig. 1 (*cont.*)
phate; F6P, fructose 6-phosphate; F1,6BP, fructose 1,6-bisphosphate; Triose-P, the equilibrium mixture of glyceraldehyde 3-phosphate and dihyroxyacetone phosphate; 3PGA, 3-phosphoglycerate; 2PGA, 2-phosphoglycerate; PEP, phosphoenol pyruvate. Reproduced from (Dennis, 1989) with permission of the American Society of Plant Physiology.

is powerfully activated by fructose 2,6-bisphosphate (Fru-2,6-P_2). In contrast to animals, plant PFK is unaffected by Fru-2,6-P_2. PFK occurs as cytosolic and plastid forms, whereas PFP has so far been found only in the cytosol (MacDonald & Priess, 1986). PFK catalyses a reaction that is irreversible under physiological conditions whereas the PFP reaction is reversible even though the standard free energy change of the two reactions is similar.

The function of cytosolic PFK appears to be the production of carbon skeletons, via glycolysis, for transport into the mitochondrion for energy metabolism. The level of PFK correlates well with energy metabolism, for example in the spadix of the arum lily when glycolysis increases rapidly during thermogenesis (ap Rees, Green & Wilson, 1985). The chloroplast form of PFK is involved in starch degradation although it can also be found in plastids that do not contain starch (Garland & Dennis, 1980). The function of PFP has not been fully clarified although it could serve different functions in different tissues. The reaction catalyzed by PFP appears to be at equilibrium (Stitt, 1989). PFP activity and concentration are higher in young metabolically active tissues (Botha, de Vries & Small, 1989) and it has been suggested that it is involved in facilitating carbon flow for biosynthetic reactions via the equilibration of the hexose and triose phosphate pools (Dennis & Greyson, 1987), a function which might be impossible for PFK because of the tight regulation of the latter enzyme (Garland & Dennis, 1980). Fru-2,6-P_2 may serve as an indicator of the metabolic state of the tissue, increasing under metabolically active conditions and consequently stimulating PFP (Dennis & Greyson, 1987). It has also been suggested that, during the import of sucrose, PFP functions in the reverse direction and generates pyrophosphate required for the uridine diphosphoglucose pyrophosphorylase reaction when sucrose synthase is involved in sucrose catabolism (ap Rees et al., 1985). Since the reaction is at equilibrium, it may serve to equilibrate all its substrates and products and could accomplish both the above functions depending on the metabolic demands of the tissue (Stitt, 1989).

PFK and PFP catalyze distinct reactions and hence they cannot be considered isozymes. However, they have closely related functions and will be considered together in this article.

The ATP-dependent phosphofructokinase

Plastid and cytosolic forms of PFK with distinct properties were first demonstrated in spinach leaves (Kelly & Latzko, 1977). In non-photosynthetic tissue, such as the developing endosperm of the castor plant, leucoplast and cytosolic PFK have been isolated and separated by ion exchange chromatography (Garland & Dennis, 1980). Cytosolic and plastid forms of PFK have also been described for the enzyme from cucumber cotyledons (Cawood, Botha & Small, 1988) and *Phaseolus vulgaris* seeds (Botha & Small, 1987). In potato tubers, four forms of PFK have been found in which the subunit relative molecular mass ranged from 46·3–53·0 kDa (Kruger, Hammond & Burrel, 1988). The cellular location of these various forms has not been determined.

The cytosolic and leucoplast forms of PFK from the castor endosperm have very different physical and kinetic properties (Garland & Dennis, 1980). The plastid form is relatively stable and has been purified to homogeneity (Knowles, Greyson & Dennis, 1990). The enzyme is a homotetramer with a native M_r of 220,000 and a subunit M_r of 57,000. It shows complex regulatory kinetics (Garland & Dennis, 1980). The saturation kinetics for one of the substrates, ATP, are hyperbolic at lower concentrations but higher concentrations of ATP are inhibitory. This inhibition is reversed by the second substrate, Fru-6-P, so that the actual inhibition curves for ATP are different at each Fru-6-P level. The ATP inhibition is also reversed by inorganic phosphate (P_i). The second substrate, Fru-6-P, has hyperbolic kinetics at pH 8.0 but these change to sigmoidal kinetics at pH 7.0. At pH 7.0, P_i will activate the enzyme and changes the saturation curve to hyperbolic. In contrast, PEP at low concentrations (Garland & Dennis, 1980) makes the curves more sigmoidal and hence is a very effective inhibitor of the enzyme. Other glycolytic intermediates will inhibit the enzyme but not as effectively as PEP. The significance of this regulation in non-photosynthetic plastids is not clear. Obviously, the ratio of P_i to PEP is the main regulatory influence. It is possible that the enzyme in leucoplasts is identical with that found in chloroplasts, retaining the same regulatory characteristics, but these may not have any metabolic significance in leucoplasts. More likely, the regulatory properties are important but their significance will not be understood until the mechanism and nature of the uptake of carbon intermediates is known.

The cytosolic form of PFK is very unstable which has so far made it impossible to purify from developing castor endosperm and its native and subunit molecular weights are not known. In contrast to the leucoplast enzyme, the cytosolic form has hyperbolic kinetics with both substrates at pH 8·0 and pH 7·0 (Garland & Dennis, 1980). Hence, the regulatory properties are not as pronounced when it is compared with the plastid enzyme (Garland & Dennis, 1980). This is surprising since it might have been expected that the flow of carbon down the glycolytic pathway in the cytosol would be subject to tight regulation as it is in other organisms. The enzyme is inhibited by PEP and other glycolytic intermediates and this inhibition is reversed by P_i so that a similar if less effective control is exerted. ATP is not inhibitory and, unlike animals, ADP does not stimulate the activity. The control of the enzyme is therefore mainly by the PEP/P$_i$ ratio. This ratio is an indication of the need for carbon intermediates either in the cytosol for transport into the mitochondrion or for carbon import into the leucoplast. Hence, in plants, the flow of carbon down the glycolytic pathway via PFK is determined primarily by carbon requirements rather than directly by energy needs.

A question that has not been resolved is whether the enzymes involved in basic metabolism such as the glycolytic pathway are the same or different in the various tissues of the plant. In animals, there are often organ-specific enzymes which might be quite different in, for example, the liver and muscle. In plants there are certainly organ-specific promotors such as those found in seeds and tubers. Since the ratio of the cytosolic to plastid form of the glycolytic enzymes can vary considerably in different tissues, it might be suspected that this is achieved through the activation of different genes in each tissue. However, it could equally well occur through differential activation of the same gene. This will not be resolved until the structure of the various genes has been elucidated and the properties of the enzymes in each tissue determined.

In order to study the problem of organ-specific enzymes, the cytosolic and leucoplast forms of PFK in the developing castor endosperm have been compared with PFK from the cytosol and chloroplast of castor leaves (Knowles et al., 1990). The two leaf forms of PFK can be separated in a very similar manner to the endosperm forms (Knowles et al., 1990). The physical and kinetic properties of the chloroplast and cytosolic forms from the leaf are very similar to those from their counterparts in the endosperm. The regulatory properties are also similar except that the endosperm form of the plastid enzyme seems somewhat more sensitive to inhibition by glycolytic intermediates. It is not known whether this difference stems from tissue-specific expression of separate genes, post-translational modifications, or if it is simply an effect of extraction from different tissues. An antibody raised against the 50 kDa subunit of PFK from potato (Kruger et al., 1988) will effectively immunoremove both the leucoplast and the chloroplast PFKs to the same extent, suggesting that they are immunologically very similar if not identical (Knowles et al., 1990). This antibody has no effect on the cytosolic form of PFK from either source suggesting that the cytosolic and plastid forms of PFK are quite distinct proteins as is the case in developing cucumber seeds (Cawood et al., 1988). Although leucoplast and chloroplast enzymes, however, appear to be identical immunologically and share similar kinetic and regulatory properties, the differences between the cytosolic and plastid enzymes suggest that they may be isozymes.

The pyrophosphate-dependent phosphofructokinase

PFP was first shown to be present in plants in 1979 (Carnal & Black, 1979), but its importance was not realized until it was shown to be activated by Fru-2,6-P_2 (Sabularse & Anderson, 1981). It is found in all higher plants and is most active in young tissues (Botha *et al.*, 1989). Several roles have been assigned to it as described above. Since PFP has distinct substrate requirements it is not an isozyme of PFK but is included in this chapter since it clearly plays a role that is closely associated with PFK. It was suggested that PFP and PFK could be interconverted (Balogh, Wong & Buchanan, 1984) but this is now known not to be the case (Kruger & Dennis, 1985). The function and significance of this enzyme have been described elsewhere (ap Rees *et al.*, 1985; Stitt, 1989) and will not be discussed here. However, of great interest is the relationship of PFK to PFP, especially from an evolutionary standpoint.

PFP occurs in the potato tuber as a heterotetramer with a native M_r of 265 kDa. It is composed of two α and two β subunits (Kruger & Dennis, 1987) with subunit M_r values of 65,000 and 60,000 respectively. The tetramer can be dissociated into a dimer by high concentrations of PP_i; reassociation is favoured by Fru-2,6-P_2. The significance of this interconversion *in vivo* is not known since non-physiological concentrations of either PP_i or Fru-2,6-P_2 are required. There is evidence that the active site of the enzyme is on the β subunit (Kruger & Dennis, 1987) whilst the α subunit may be involved in the activation of the enzyme.

Antibodies have been raised to the separated α and β subunits and it has been shown that they are immunologically distinct (Kruger & Dennis, 1987). These antibodies have been used to isolate cDNA clones from a castor endosperm λgt11 library and full-length clones from a potato λZAP library (Carlisle *et al.*, 1990). These clones have been sequenced and from this the amino acid sequence has been derived. The clones for each subunit from castor and potato are very similar with 83% and 87% of the amino acids being identical for the α and β subunits, respectively. Many of the non-identical amino acids represent conservative changes. When the derived amino acid sequence of the α and β clones are compared it is found that 40% of the amino acids are the same with further 20% of changes being conservative. Hence, although there is no cross reaction of antibodies between the α and β subunits there is a considerable amount of similarity at the amino acid level.

When the derived amino acid sequence of PFP is compared with the amino acid sequence of PFK from other organisms little similarity is found, confirming that these enzymes are distinct proteins. However, a detailed comparison of the PFP sequences with the PFK-1 sequence of the enzyme from *E. coli* (Shirakihara & Evans, 1988) reveals that, although the overall homology is low, eight of the eleven amino acids involved in Fru-6-P, Fru-1,6-P_2 binding can be found in the β subunit of PFP each being flanked by identical or similar amino acids in the two enzymes. These sites are also conserved to a large extent in the α subunit of PFP except that an aspartate essential to the catalytic activity of PFK has been replaced in this subunit by an asparagine which would likely render it catalytically inactive. This evidence is consistent with the suggestion that the catalytic site of PFP is on the β subunit. In addition, the β subunit has been reported to be the only one present in some tissues (Yan & Tao, 1984). Detailed examination of the ATP/ADP substrate binding site of *E. coli* PFK-1 reveals the phosphate domain of this site may be conserved in PFP and could possibly account for the binding of pyrophosphate. Sites involved in the regulatory activity of PFK have not been found in either subunit of PFP.

These data indicate that the subunits of PFP are related and probably derived from a gene duplication. The sequence data are consistent with the β subunit having the active site. In spite of their similarities, there is no cross reactivity of antibodies made to the individual subunits (Kruger & Dennis, 1987). The similarities in the substrate binding sites between PFK and PFP indicate that the proteins are distantly related and may have arisen from a common ancestral gene. In most cases, antibodies raised against either PFP or PFK do not recognize the other enzyme (Kruger & Hammond, 1988). However, the presence of similar sequences at the substrate binding sites could account for the weak cross reaction that has been described for *E. coli* PFK-1 antibodies and the β subunit of PFP (Yuan, Kwiatkowska & Kemp, 1988).

Phosphoglyceromutase

Phosphoglyceromutase (PGM) is of particular interest in the glycolytic pathway in plants for several reasons. It is present in the cytosol and plastids in some plants (Botha & Dennis, 1986) but the plastid form is not always found (Stitt & ap Rees, 1979). In chloroplasts, 3-phosphoglycerate is the metabolite that is formed in carbon fixation during photosynthesis and the absence of this enzyme may prevent the draining of photosynthate into the plastid glycolytic pathway which could inhibit photosynthesis. Of particular interest, however, is the fact that both the plastid and cytosolic enzymes from plants do not require 2,3-bisphosphoglycerate as a cofactor which is the case with the enzyme from animals and many other organisms.

The cytosolic and plastid forms of PGM are very similar enzymes that can be separated by ion filtration chromatography but only under precise conditions of pH (Miernyk & Dennis, 1982). The percentage of PGM activity that is found in the leucoplast of the castor endosperm varies between 5 and 9% which, at the lower end, is difficult to distinguish from cytosolic contamination. Both isozymes are monomeric, with M_r values of approximately 63,000. The plastid form of PGM is more heat labile than the cytosolic form. Other kinetic and physical properties are very similar but sufficiently different to indicate that they are distinct proteins (Botha & Dennis, 1986). Antibodies have been raised to the purified cytosolic form of PGM from castor endosperm (Botha & Dennis, 1987). There is cross reaction of this antibody with the leucoplast form, suggesting that the two are very similar immunologically. However, because the two forms are difficult to separate completely, this cross-reactivity could also result from leucoplast PGM contamination of the original antigen to which the antibody was raised. PGM has been purified to homogeneity from wheat (Leadley *et al.*, 1977) and rice germ (Fernandez & Grisdia, 1960) but the intracellular location of the enzyme was not determined. The physical and kinetic properties of the enzymes from these sources are very similar to those described for the castor enzymes.

The relationship between the plant, 2,3-bisphosphoglycerate-independent enzyme and the dependent enzyme will only become clear when sequence data of the plant proteins are available and can be compared with the known sequences of the enzyme from yeast and mammals, both of which require the cofactor. The antibodies described above are being used to isolate cDNA clones for the castor endosperm PGM and a putative clone has been isolated which shows limited homology with the yeast and *Neurospora* enzymes, the latter of which is also a cofactor independent form. However, this comparison is preliminary and full-length clones need to be isolated before the relationship beween these enzymes can be elucidated.

Enolase

The leucoplast and cytosolic forms of enolase from castor endosperm can be easily separated by ion filtration chromatography (Miernyk & Dennis, 1982) and they have been partially purified (Miernyk & Dennis, 1984). Because of the ease of separation of the two forms, enolase has been used to determine the magnitude of the glycolytic pathway in various parts of the castor oil plant (Miernyk & Dennis, unpublished). It was found that the percentage of plastid enolase increased during seed development until it reached approximately 30% of the total activity at the time of maximum oil biosynthesis. As the seed matured and dried, this percentage fell again. During germination, plastid enolase was also found, but increased to a maximum of only 10% of the total activity. In mature leaves, no chloroplast enolase was present but it was found at a level of 7% in immature leaves. About 7% of enolase activity was attributed to the plastid form in roots. Enolase has also been found to be absent from spinach chloroplasts. These data illustrate that there is a tight developmental and tissue specific control of the level of plastid glycolytic enzymes. How this control is exerted is not known.

Both forms of castor enolase have properties that are typical of enolases from other sources but there are clear differences between them, other than their difference in charge, which makes them easy to separate (Miernyk & Dennis, 1984). The enzymes have similar pH optima for the forward reaction but differ markedly in the optimum for the reverse reaction. The Michaelis constants for the substrates 2-PGA and PEP are lower for the plastid form of enolase. One surprising result was that antibodies raised against yeast enolase cross-reacted only with the leucoplast form and showed no reaction with the cytosolic enzyme, indicating that the two forms are immunologically different (Miernyk & Dennis, 1984). These antibodies did not cross-react with either *E. coli* or rabbit muscle enolases.

Pyruvate kinase

Pyruvate kinase (PK) occupies a major branch point in metabolism and hence should play a central role in the control of the glycolytic pathway, especially in plants. In animals, the enzyme has regulatory properties but none of any consequence have been described for the enzyme in plants. The animal enzyme is also modified by phosphorylation and again this has not been demonstrated in plants.

As with the other glycolytic enzymes in the castor endosperm, a leucoplast and cytosolic form of PK have been isolated (Ireland *et al.*, 1980). The plastid form of PK is heat labile which has made it difficult to purify. In contrast, the cytosolic form is much more stable and has been purified to homogeneity (Plaxton, 1988). The pH optimum for the cytosolic castor PK is approximately 7·0 whereas that of the plastid form is 8·0 (Ireland *et al.*, 1980). The affinity for substrates is much higher with the cytosolic form than it is with the plastid form. Both enzymes have an ordered sequential reaction mechanism but the order of product release in the two forms is different. In the cytosolic form of PK, ATP is the last product to be released whereas pyruvate is the last product to be released from the leucoplast form. ADP is the first substrate to bind in both forms of PK. For the leucoplast enzyme, this means that pyruvate is competitive with ADP binding (Ireland *et al.*, 1980). Pyruvate, the product of the leucoplast glycolytic pathway in the developing endosperm of the castor plant, is used for fatty acid biosynthesis. A build-up of pyruvate would indicate that the production of carbon precursors exceeded the rate of fatty acid biosynthesis. This build-up would inhibit PK via competition with ADP and would prevent the flow of carbon down the leucoplast glycolytic pathway (Ireland *et al.*, 1980). In contrast, ATP is the final product released from cytosolic PK which means that it competes with the binding of ADP and may play a role in the control of energy metabolism.

Homogeneous cytosolic PK has been prepared from germinating castor oil seed (Plaxton, 1988). It appears to be a heterotetramer of two subunits with M_r values of 56,000 and 57,000. These subunits are distinct but related proteins. In contrast, the cytosolic form of PK from developing castor seed is a homotetramer subunits of 56,000, as is the enzyme from leaves (Plaxton, 1989). The significance of these differences are not known but is probably related to differences in the regulatory properties of the enzymes. Antibodies raised against the cytosolic form of PK recognize cytosolic PKs from other tissues and other plants but do not recognize the plastid form of the enzyme.

The green alga *Selanastrum minutum* has cytosolic and chloroplast forms of PK which can be separated (Lin, Turpin & Plaxton, 1989). The cytosolic form has properties that are similar to the cytosolic form of the castor enzyme and the antibodies to that enzyme cross-react with that from *Selanastrum*. In contrast, the chloroplast form appears to be quite different from any known PK. This enzyme has been purified to homogeneity and antibodies raised to it (Knowles, Dennis and Plaxton, submitted). These antibodies do not recognize the cytosolic enzyme from *Selanastrum* or other sources. Surprisingly, these antibodies do not recognize the castor leucoplast isozyme, suggesting that it is quite different from either higher plant enzyme. However, the antibodies do recognize PK from *Baccillus subtilis*. In addition, unlike other PKs the *Selanastrum* chloroplast PK is a large monomeric protein with a M_r of 235,000. Clones for the cytosolic and chloroplast forms of PK are being isolated from a *Selanastrum* cDNA library to determine the relationship between these PKs and those from other sources.

Clones for the cytosolic form of PK have been isolated from potato and developing castor endosperm cDNA libraries (Blakeley, Plaxton & Dennis, 1990). These clones have been sequenced and the derived amino acid sequence data available so far has indicated that the plant cytosolic enzyme is very similar to pyruvate kinases from other sources. The ADP and PEP binding domains are highly conserved in all these PKs. However, there is no sequence that resembles the phosphorylation site of the yeast and mammalian enzymes consistent with there being no report of phosphorylation of the plant enzyme. An interesting feature of these castor PK clones is that they possess a 3' untranslated region that varies in length from about 1100 to 2000 bp. Similar, long and variable 3' untranslated regions are also a feature of the enzyme from mammals where it has been suggested that this region is involved in the stability of the mRNA and hence its level of expression (Marie *et al.*, 1986).

Originally it was thought that all the PKs found in rats existed as true isomeric forms. More recently it has been found that a single gene codes for the R and L forms of rat PK (Noguchi *et al.*, 1987). The two enzymes are formed by a differential expression of the 12 exons that make up the gene for PK through the use of different promoters (Noguchi *et al.*, 1987). Furthermore, differential splicing of exons generates the M_1 and M_2 forms of the muscle enzyme (Noguchi *et al.*, 1986). Initial PCR data on genomic DNA from castor and potato plants has suggested that introns are also found in the gene for cytosolic PK (Cole, Blakeley & Dennis, unpublished). It is possible, therefore, that the regulation of expression of cytosolic PK is also mediated through tissue or developmental control of transcription of the PK gene.

Future directions

There is still a great deal to be learned about the structure and function of the enzymes that make up the glycolytic pathway in plants. In particular, the relationship between the plastid and cytosolic forms of these enzymes needs to be elucidated. This will require more information about the physical and kinetic differences between the isoforms after they have been purified to homogeneity.

As yet, it is not known whether any of these enzymes are encoded by multigene families and whether there are tissue-specific promotors for them. Little is known about post-transcriptional or post-translational modifications of gene transcripts and whether this plays any role in the differences that have been found in the isoforms. The differences in the genes that encode the plastid and cytosolic forms of the glycolytic enzymes need to be elucidated, if indeed the plastid and cytosolic enzymes are true isozymes. Finally the mechanism by which these genes are regulated to enable the ratio of the two forms to be controlled thereby affecting the metabolism of specific tissues has to be determined.

References

ap Rees, T., Green, J. H. & Wilson, P. M. (1985). Pyrophosphate:fructose 6-phosphate 1-phosphotransferase and glycolysis in non-photosynthetic tissues of higher plants. *Biochem. J.* **227**, 299–304.

Balogh, A., Wong, J. H., Buchanan, B. B. (1984). Metabolite-mediated interconversion of PFP/PFK: a regulatory mechanism to direct cytosolic carbon flux. *Plant Physiol.* **75**, S–53.

Blakeley, S. D., Plaxton, W. C. & Dennis, D. T. (1990). Cloning and characterisation of a cDNA for the cytosolic isozyme of plant pyruvate kinase: the relationship between the plant and non-plant enzyme. *Plant. Mol. Biol.* (In press.)

Botha, F. C. & Dennis, D. T. (1986). Isozymes of phosphoglyceromutase from the developing endosperm of *Ricinus communis*: isolation and kinetic properties. *Arch. Biochem. Biophys.* **245**, 96–103.

Botha, F. C. & Dennis, D. T. (1987). Phosphoglyceromutase activity and concentration in the endosperm of developing and germinating *Ricinus communis* seeds. *Can. J. Bot.* **65**, 1908–12.

Botha, F. C., de Vries, C. & Small, J. G. C. (1989). Changes in the activity and concentration of the pyrophosphate dependent phosphofructokinase during germination of *Citrullus lanatus* seeds. *Plant Physiol. Biochem.* **27**, 75–80.

Botha, F. C. & Small, J. G. C. (1987). Comparison of the activities and some properties of pyrophosphate and ATP-dependent fructose 6-phosphate 1-phosphotransferase of *Phaseolus vulgaris*. *Plant Physiol.* **83**; 772–7.

Boyle, S. A., Hemmingsen, S. M. & Dennis, D. T. (1986). Uptake and processing of the precursor to the small subunit of ribulose 1,5-bisphosphate carboxylase by leucoplasts from the endosperm of developing castor oil seeds. *Plant Physiol.* **81**, 817–22.

Boyle, S. A., Hemmingsen, S. M. & Dennis, D. T. (1989). Energy requirement for the import of protein into plastids from developing endosperm of *Ricinus communis L. Plant Physiol.* (In press.)

Carlisle, S. A., Blakeley, S. D., Hemmingsen, S. M., Trevanion, S. J., Hiyoshi, T., Kruger, N. J. & Dennis, D. T. (1990). Pyrophosphate dependent phosphofructokinase: conservation of protein sequence between the alpha and beta subunits and with the ATP dependent phosphofructokinase. *J. Biol. Chem.* (In press.)

Carnal, N. W. & Black, C. C. (1979). Pyrophosphate-dependent 6-phosphofructokinase. A new glycolytic enzyme in pineapple leaves. *Biochem. Biophys. Res. Commun.* **86**, 20–6.

Carnal, N. W. & Black, C. C. (1983). Phosphofructokinase activities in photosynthetic organisms: the occurrence of pyrophosphate-dependent 6-phosphofructokinase in plants and algae. *Plant Physiol.* **71**, 150–5.

Cawood, M. E., Botha, F. C. & Small, J. G. (1988). Molecular properties of the ATP: D-fructose 6-phosphate 1-phosphotransferase isoenzymes from *Cucumis sativus*. *Plant Cell Physiol.* **29**, 195–9.

Deng, X. & Gruissem, W. (1987). Control of plastid gene expression during development: the limited role of transcriptional regulation. *Cell* **49**, 379–87.

Dennis, D. T., (1989). Fatty acid biosynthesis in plastids. In *Physiology, Biochemistry, and Genetics of Nongreen Plastids*, ed, C. D. Boyer, J. C. Shannon & R. C. Hardison, Proceedings of the Fourth Annual Penn State Symposium, Amer. Soc. of Plant Physiol.

Dennis, D. T., & Greyson, M. F. (1987). Fructose 6-phosphate metabolism in plants. *Physiol. Plant.* **69**, 395–404.

Dennis, D. T., Hekman, W. E., Thomson, A., Ireland, R. J., Botha, F. C. & Kruger, N. J. (1985). Compartmentation of glycolytic enzymes in plants. In *Regulation of Carbon Partitioning in Photosynthetic Tissues*, eds. R. L. Heath & J. Preiss, pp. 127–46. Amer. Soc. Plant Physiol. Symp., Riverside Calif.

Dennis, D. T. & Miernyk, J. A. (1982). Compartmentation of non-photosynthetic carbohydrate metabolism. *Ann. Rev. Plant Physiol.* **33**, 17–50.

Fernandez, M. & Grisdia, S. (1960). Phosphoglyceric acid mutase activity without added 2,3-diphosphoglycerate in preparations purified from rice germ. *J. Biol. Chem.* **235**, 2188–90.

Garland, W. J. & Dennis, D. T. (1980). Plastid and cytosolic phosphofructokinases from the developing endosperm of *Ricinus communis*. I. Separation, purification, and initial characterization of the isozymes. *Arch. Biochem. Biophys.* **204**, 302–9.

Garland, W. J. & Dennis, D. T. (1980). Plastid and cytosolic phosphofructokinases from the developing endosperm of *Ricinus communis*. II. Comparison of the kinetic and regulatory properties of the isozymes. *Arch. Biochem. Biophys.* **204**, 310–37.

Highfield, P. E. & Ellis, R. J. (1978). Synthesis and transport of the small subunit of chloroplast ribulose bisphosphate carboxylase. *Nature* **271**, 420–4.

Ireland, R. J., DeLuca, V. & Dennis, D. T. (1980). Characterization and kinetics of pyruvate kinase from developing castor oil seed endosperm. *Plant Physiol.* **65**, 1188–93.

Ireland, R. J. & Dennis, D. T. (1980). Isozymes of the glycolytic and pentose phosphate pathway in storage tissues of different oilseeds. *Planta* **149**, 476–9.

Journet, E.-P. & Douce, R. (1985). Enzymic capacities of purified cauliflower bud plastids for lipid synthesis and carbohydrate metabolism. *Plant Physiol.* **79**, 458–67.

Kelly, G. J. & Latzko, E. (1977). Chloroplast phosphofructokinase. I. Proof of phosphofructokinase activity in chloroplasts. *Plant Physiol.* **60**, 290–4.

Kleinig, H. & Liedvogel, B. (1980). Fatty acid synthesis by isolated chromoplasts from the daffodil. Energy sources and distribution patterns of the acids. *Planta* **150**, 166–9.

Knowles, V. L., Dennis & Plaxton, W. C. (1990). Purification of a novel pyruvate kinase from a green alga. *FEBS Letters*, **259**, 130–2.

Knowles, V. L., Greyson, M. F. & Dennis, D. T. (1990). Characterization of ATP dependent fructose 6-phosphate 1-phosphotransferase from leaf and endosperm tissue of *Ricinus communis*. *Plant Physiol.* **92**, 155–9.

Kruger, N. J. & Dennis, D. T. (1985). A source of apparent pyrophosphate: fructose 6-phosphate phosphotransferase activity in rabbit muscle phosphofructokinase. *Biochem. Biophys. Res. Commum.* **126**, 320–6.

Kruger, N. J. & Dennis, D. T. (1987). Molecular properties of pyrophosphate:fructose 6-phosphate phosphotransferase from potato tuber. *Arch. Biochem. Biophys.* **256**, 273–9.

Kruger, N. J. & Hammond, J. B. W. (1988). Molecular comparison of pyrophosphate- and ATP-dependent fructose 6-phosphate 1-phosphotransferase from potato tuber. *Plant Physiol.* **86**, 645–8.

Kruger, N. J., Hammond, J. B. W. & Burrel, M. M. (1988). Molecular characterization of four forms of phosphofructokinase purified from potato tuber. *Arch. Biochem. Biophys.* **267**, 690–700.

Leadley, P. F., Breathnach, R., Gatehouse, J. A., Johnson, P. E. & Knowles, J. R. (1977). Phosphoglycerate mutase from wheat germ: isolation, crystallization and properties. *Biochemistry* **16**, 3050–3.

Lin, M., Turpin, D. H. & Plaxton, W. C. (1989). Pyruvate kinase isozymes from the green alga *Selanastrum minutum*. I Purification and physical and immunological characterization. *Arch. Biochem. Biophys.* **269**, 219–27.

Macdonald, F. D. & Priess, J. (1986). The subcellular location and characteristics of pyrophosphate-fructose-6-phosphate 1-phosphotransferase from suspension-culture cells of soybean. *Planta* **167**, 240–5.

Marie, J., Simon, M.-P., Lone. Y.-C., Cognet, M. & Kahn, A. (1986). Tissue specific heterogeneity of the 3′-untranslated region of L-type pyruvate kinase mRNAs. *Eur. J. Biochem.* **158**, 33–41.

Miernyk, J. A. & Dennis, D. T. (1982). Isozymes of the glycolytic enzymes in endosperm from developing castor oil seeds. *Plant Physiol.* **69**, 825–8.

Miernyk, J. A. & Dennis, D. T. (1983). The incorporation of glycolytic intermediates into lipids by plastids isolated from the developing endosperm of castor oil seeds (*Ricinus communis* L.). *J. Exp. Bot.* **34**, 712–18.

Miernyk, J. A. & Dennis, D. T. (1984). Enolase isozymes from *Ricinus communis*: partial purification and characterization of the isozymes. *Arch. Biochem. Biophys.* **233**, 643–51.

Noguchi, T., Inoue, H. & Tanaka, T. (1986). The M_1- and M_2-type isozymes of rat pyruvate kinase are produced from the same gene by alternative RNA spicing. *J. Biol. Chem.* **261**, 13807–12.

Noguchi, T., Yamada, K., Inoue, H., Matsuda, T. & Tanaka, T. (1987). The L- and R-type isozymes of rat pyruvate kinase are produced from a single gene by use of different promoters. *J. Biol. Chem.* **262**, 14366–71.

Ohlrogge, J. B., Kuhn, D. N. & Stumpf, P. K. (1979). Subcellular localization of acyl carrier protein in leaf protoplasts of *Spinacia oleracea*. *Proc. Natn. Acad. Sci. USA* **76**, 1194–8.

Plaxton, W. C. (1988). Purification of pyruvate kinase from germinating castor bean endosperm. *Plant Physiol.* **86**, 1064–9.

Plaxton, W. C. (1989). Molecular and immunological characterization of plastid and cytosolic pyruvate kinase isozymes from castor oil plant endosperm and leaf. *Eur. J. Biochem.* **181**, 443–51.

Reid, E. E. Thompson, P., Lyttle, C. R. & Dennis, D. T. (1977). Pyruvate dehydrogenase complex from higher plant mitochondria and proplastids. *Plant Physiol.* **59**, 842–8.

Sabularse, D. C. & Anderson, R. L. (1981). D-fructose, 2,6-bisphosphate: a naturally occurring activator for inorganic pyrophosphate D-fructose 6-phosphate 1-phosphotransferase in plants. *Biochem. Biophys. Res. Commun.* **103**, 848–55.

Shirakihara, Y. & Evans, P. R. (1988). Crystal structure of the complex of phosphofructokinase from *Escherichia coli* with its reaction products. *J. Molec. Biol.* **204**, 973–94.

Simcox, P. D., Reid, E. E., Canvin, D. T. & Dennis, D. T. (1977). Enzymes of the glycolytic and pentose phosphate pathway in proplastids from the developing endosperm of *Ricinus communis*. L. *Plant Physiol.* **59**, 1128–32.

Stitt, M. (1989). Product inhibition of potato tuber pyrophosphate:fructose 6-phosphate phosphotransferase by phosphate and pyrophosphate. *Plant Physiol.* **89**, 628–33.

Stitt, M. & ap Rees, T. (1979). Capacities of pea chloroplasts to catalyze the oxidative pentose phosphate pathway and glycolysis. *Phytochemistry* **18**, 1905–11.

Weeden, N. F. & Gottlieb, L. D. (1980). The genetics of chloroplast enzymes. *J. Hered.* **71**, 392–6.

Yan, T.-F. J. & Tao, M. (1984). Multiple forms of pyrophosphate:D-fructose-6-phosphate 1-phosphotransferase from wheat seedlings. *J. Biol. Chem.* **259**, 5087–92.

Yuan, X.-H., Kwiatkowska, D. & Kemp, R. G. (1988). Inorganic pyrophosphate:fructose 6-phosphate 1-phosphotransferase of the potato tuber is related to the major ATP-dependent phosphofructokinase of *E. coli*. *Biochem. Biophys. Res. Commun.* **154**, 113–17.

Zilkey, B. F. & Canvin, D. T. (1972). Localization of oleic acid biosynthesis enzymes in the proplastids of developing castor endosperm. *Can. J. Bot.* **50**, 323–6.

T. ᴀᴘ REES, T. G. ENTWISTLE AND
J. E. DANCER

Interconversion of C-6 and C-3 sugar phosphates in non-photosynthetic cells of plants

We intend to discuss the importance, extent, and mechanisms of the interconversion of hexose monophosphates and triose phosphates in the non-photosynthetic, non-gluconeogenic cells of higher plants.

Importance

The importance of our topic is twofold. First, conversion of hexose monophosphate to triose phosphate is generally regarded as committing the cell's supply of carbon and energy to respiration. Such commitment is vital to the plant's economy as respiration is a major drain on its resources. Classically, the entry of hexose monophosphate into respiration has been seen as an irreversible process catalysed by phosphofructokinase [PFK: EC 2·7·1·11] in the cytosol. This view has been complicated by the discoveries that glycolysis occurs in the plastid as well as the cytosol (ap Rees, 1985), and that the cytosol of plants contains pyrophosphate:fructose-6-phosphate 1-phosphotransferase [PFP: EC 2·7·1·90] (ap Rees, 1988). This enzyme catalyses

$$PP_i + Fru\text{-}6\text{-}P \;\rightleftharpoons\; P_i + Fru\text{-}1,6\text{-}P_2$$

The second reason for considering our subject is that it is central to the interconversion of the two major storage compounds of plants, sucrose and starch. The non-photosynthetic cells of plants receive their substrate largely as sucrose. Available evidence suggests that it is either delivered as sucrose via the plasmadesmata or absorbed from the apoplast as sucrose or, more rarely, hexose (Ho, 1988). Starch synthesis is confined to the plastid and thus depends upon transport of substrate from the cytosol. If, as comparative biology might suggest, this transport occurs as triose phosphate via the phosphate translocator, then sucrose arriving in the starch-storing cell will have to be converted to triose phosphate in the cytosol for entry into the plastid, which must then be able to convert triose phosphate to hexose monophosphate for starch synthesis. A num-

95

Table 1. *Maximum catalytic activities and distribution of enzymes of sugar phosphate metabolism in the developing endosperm of wheat. Distribution is from Entwistle & ap Rees (1988): activities were determined by applying the same assays to extracts of endosperm made in 40 mM glycylglycine (pH 7·4) and centrifuged at 100,000 × g for 30 min*

Enzyme	Activity (μmol min^{-1} g^{-1} fresh wt.)	Percentage of activity in amyloplasts
Phosphoglucomutase	1·92	11
Glc-6-P isomerase	16·50	25
PFK	0·31	23
Aldolase	1·39	26
Triosephosphate isomerase	342·30	19
PFP	1·26	0

ber of starch-storing cells convert starch to sucrose, e.g. potato tubers and the cotyledons of germinating peas. If the plastid envelope in such cells remains intact then some product of starch breakdown will have to move from the plastid to the cytosol to support sucrose synthesis. If this transport occurs as a 3-C compound then the cytosol will have to be able to convert triose phosphate to hexose monophosphate.

The central importance of the relationship Fru-6-P: Fru-1,6-P$_2$

Understanding the relationship between C-6 and C-3 sugar phosphates requires knowledge of the maximum catalytic activities of the appropriate enzymes and the concentrations of the relevant metabolites, in the cytosol and the plastid. Such information is sparse, largely because no one has managed to measure metabolites in the plastid and in the cytosol of non-photosynthetic cells. The data in Tables 1 and 2 are the best that we can muster. The substrate measurements are from photosynthetic cells. Their extrapolation to non-photosynthetic cells is to some extent justified by the fact that each of the conclusions to be drawn from Table 2 is supported by authenticated measurements of the total tissue contents of the metabolites in a range of non-photosynthetic tissues of plants (Dixon & ap Rees, 1980; ap Rees, Green & Wilson, 1985; Dancer & ap Rees, 1989b).

Table 2. *Estimates of metabolite concentrations in illuminated leaves. Volumes of cytosol and chloroplasts are taken as 20 μl per mg chlorophyll. Data are from* Spinacia *except that ATP and ADP in the cytosol refer to wheat*

Compound	Concentration in cytosol (mM)	Reference	Concentration in chloroplast (mM)	Reference
Glc-1-P	0·6	Stitt *et al.* (1987)	0·25	Wirtz *et al.* (1980)
Glc-6-P	4·0		1·20	
Fru-6-P	1·4		0·55	
Fru-1,6-P$_2$	0·1		0·90	
D-Glyceraldehyde-3-P	0·05		0·022	Dietz & Heber (1984)
Dihydroxyacetone-P	0·3		0·48	
P$_i$	0·3		6·95	Stitt *et al.* (1980)
PP$_i$	10·0		—	
ATP	1·8	Stitt *et al.* (1984)	0·75	Wirtz *et al.* (1980)
ADP	0·3		0·55	

We have calculated, for cytosol and plastid, the mass-action ratios for each reaction (Table 3). These are the ratios of Product to Substrate in the tissues. The relatively high enzyme activities, and the lack of any marked difference between the mass action ratios and apparent equilibrium constants strongly suggest that the reactions catalysed by phosphoglucomutase, glucose-6-phosphate isomerase, aldolase and triose-phosphate isomerase are so close to equilibrium *in vivo* as to be controlled by the relative concentrations of substrates and products. Similar considerations provide sound evidence that the steps catalysed by PFK and fructose-1,6-bisphosphatase (EC 3·1·3·11, FBPase) are far from equilibrium and are highly regulated *in vivo*. These arguments hold for both plastid and cytosol, almost certainly apply to animal and microbial as well as plant cells, and pinpoint the interconversion of Fru-6-P and Fru-1,6-P$_2$ as the key to the relationship between C-6 and C-3 sugar phosphates. We now consider the properties and roles of the enzymes capable of catalysing the above interconversion.

Table 3. *Comparison of mass action ratios and apparent equilibrium constants for reactions of sugar phosphate metabolism*

			Mass action ratio	
Reaction	Reactants	Apparent K_{eq}	Cytosol	Chloroplast
Phosphoglucomutase	Glu-6-P: Glu-1-P	17·2–19·0	6·70	4·80
Glc-6-P isomerase	Fru-6-P: Glu-6-P	0·36–0·47	0·35	0·46
PFK	$\dfrac{[\text{Fru-1,6-P}_2]\,[\text{ADP}]}{[\text{Fru-6-P}]\,[\text{ATP}]}$	900–1200	0·01	1·20
Aldolase	$\dfrac{[\text{Triose phosphate}]^2 \times 22}{[\text{Fru-1,6-P}_2] \times 23^2}$	$6\cdot8 \times 10^{-5}$	$5\cdot1 \times 10^{-5}$	$1\cdot2 \times 10^{-5}$
Triosephosphate isomerase	$\dfrac{[\text{Glyceraldehyde-3-P}]}{[\text{Dihydroxyacetone-P}]}$	0·036–0·045	0·17	0·046
FBPase	$\dfrac{[\text{Fru-6-P}]\,[\text{P}_i]}{[\text{Fru-1,6-P}_2]}$	530	4·20	4·20
PFP	$\dfrac{[\text{Fru-1,6-P}_2]\,[\text{P}_i]}{[\text{Fru-6-P}]\,[\text{PP}_i]}$	3·20	2·40	—

Ratios are calculated from data in Table 2: K_{eq} from Newsholme & Start (1973).

PFK(ATP)

Many preparations of PFK have been made from plants and characterized (Copeland & Turner, 1987). In general this work revealed complex regulatory properties: positive and negative cooperativity with substrates, modulation by a range of effectors. ATP was found to be inhibitory and such inhibition was relieved by P_i but not by ADP or AMP. Marked inhibition was found with citrate and with low concentrations of the phosphoenolpyruvate (K_i, 1·3 μM) and related glycolytic intermediates. The latter was relieved by P_i and Fru-6-P, but enhanced by ATP and citrate. The picture emerged of an enzyme inhibited by ATP, phosphoenolpyruvate and citrate with P_i as the main positive effector.

Most of the above observations were made before it was apparent that distinct cytosolic and plastidic forms of PFK existed in both non-photosynthetic and photosynthetic cells of plants. Thus we can not relate the observed properties to a particular enzyme. Much of the data probably reflect the behaviour of the plastidic enzyme as this is more stable than the cytosolic form. Studies of the separate enzymes in purified form have been too few to reveal the specific characteristics of the cytosolic and

Table 4. *Activities of glycolytic enzymes and PFP during the development of the club of the spadix of* Arum maculatum

Enzyme	Stage of development	Activity (μmol min^{-1} per club)			Pre-thermogenesis	Thermo-genesis
		α	β	γ		
PFK		0·21	2·1	11·1	22·4	19·98
Phosphoglucomutase		5·40	21·0	28·0	28·7	24·31
Aldolase		0·68	4·7	13·6	17·3	10·2
Glyceraldehyde-3-P dehydrogenase		1·05	8·9	–	100·8	–
PFP		1·28	3·4	5·0	4·9	4·3
CO_2 production		0·08	0·32	–	0·78	14·3

Data from ap Rees (1977) and ap Rees, Green & Wilson (1985).

plastidic enzymes from non-photosynthetic cells of plants. Recently the situation has become even more complicated with the demonstration that different tissues of a number of varieties of potato contain not two, but four, distinct forms of PFK. Each of these forms contained different proportions of four distinct polypeptides (Kruger, Hammond & Burrell, 1988). Thus we must now face the possibility that plants as a whole may contain four forms of PFK. We urgently need to settle this issue and to characterize, and locate in the cell, the different forms.

The only established role of PFK in organisms as a whole is catalysis of the first committed step of glycolysis. There is no sound evidence that this generalization does not apply to plants and the following evidence that it does. PFK is universally distributed in plants. There is adequate evidence that it is always present in the cytosol, and each type of plastid examined in detail has been found to contain the enzyme (ap Rees, 1985). Cytosolic glycolysis is universal in plants and there is increasing evidence that at least the steps from Fru-6-P to triose phosphate also operate in the plastids (ap Rees, 1985; Emes & Bowsher, this volume). The maximum catalytic activity of PFK changes in parallel with those of other glycolytic enzymes during development, and these changes are accompanied by corresponding changes in glycolytic flux (Table 4).

We argue that the role of PFK is to provide an irreversible entry of hexose 6-phosphate into glycolysis and hence respiration, and that such entry occurs independently and simultaneously in the cytosol and the plastid. The extent to which PFK regulates such entry is more difficult to assess. The fact that the enzyme catalyses non-equilibrium reactions in the cytosol and plastid and shows regulatory properties *in vitro*, together with comparative biochemistry, strongly suggests a major contribution to glycolytic control. For a number of plant tissues we have more direct evidence that PFK is regulatory. This evidence comprises demonstrations that the amounts of Fru-6-P in tissues change in the opposite direction to the glycolytic flux when the latter is varied. This particular approach to the study of glycolytic control may not always give definitive results in plants because of the complicating presence of PFP. At present we do not know the contribution that PFP makes to glycolytic control in plants, nor, because of lack of detailed knowledge of the enzyme's properties, do we know the precise mechanism of any such control.

(PFP)

PFP is composed of two polypeptides of apparent M_r of 58,000 and 55,700 (Kruger & Hammond, 1988). The enzyme is widely, probably universally, distributed in higher plants, confined to the cytosol, specific for PP_i, requires a divalent cation and is markedly stimulated by the signal metabolite, Fru-2,6-P_2, (ap Rees, 1985). The latter decreases K_m^{app} for Fru-6-P, PP_i, and Fru-1,6-P_2. At concentrations of up to 1·5 mᴍ, P_i inhibits the forward reaction (formation of P_i), but higher concentrations of P_i inhibit both the forward and reverse reaction by decreasing the affinity for Fru-2,6-P_2 and consequently for the other three substrates. PP_i is a powerful inhibitor of the reverse reaction (Kombrink, Kruger & Beevers, 1984; Stitt, 1989).

The observation that the reaction catalysed by PFP is close to equilibrium *in vivo* makes it very difficult to identify the role of this enzyme. Four possibilities are considered: conversion of Fru-1,6-P_2 to Fru-6-P in gluconeogenesis, of Fru-6-P to Fru-1,6-P_2 in glycolysis, synthesis of PP_i, and breakdown of PP_i.

Gluconeogenesis, in which Fru-1,6-P_2 is converted to Fru-6-P in the cytosol during sucrose synthesis, occurs in photosynthetic cells and germinating fatty seeds. Both types of tissue show high activities of PFP. Both types of tissue also have a highly regulated cytosolic FBPase. There is decisive evidence that the latter makes a dominant contribution to the control of gluconeogenic flux in leaves (Stitt, Huber & Kerr, 1987), and fatty seedlings (Leegood & ap Rees, 1978; Kruger & Beevers, 1984). Cytosolic conversion of Fru-1,6-P_2 to Fru-6-P could also occur in non-photosynthetic and non-gluconeogenic tissues during the conversion of starch to sucrose if the products of starch breakdown left the plastid as triose phosphate. It is unlikely that PFP plays a major role in any of the above conversions of Fru-1,6-P_2 to Fru-6-P as high activities of PFP are found in tissues in which such conversions do not occur, and an alterna-

Table 5. *Pyrophosphate content and rate of respiration*

Tissue	Condition	CO_2 production ($\mu l\ h^{-1}\ g^{-1}/$ fresh wt)	PP_i content[a] (nmol $g^{-1}/$ fresh wt)
Club of *Arum maculatum*	Pre-thermogenesis	949	27·5±3·4
	Thermogenesis	30,000	36·2±3·4
γ-stage club of *Arum maculatum*	Untreated	—	19·9±2·6
	2,4-Dinitrophenol	—	17·6±1·7
	Aerobic	—	15·7±2·8
	Anoxic	—	14·3±2·2
Root apices of *Pisum sativum*	Untreated	461	7·0±1·0
	2,4-Dinitrophenol	623	6·7±0·5
	Aerobic	476	13·7±1·7
	Anoxic	330	13·0±1·8

[a]Values are means ± SE for at least 6 samples.
Data from ap Rees, Green & Wilson (1985) and Dancer & ap Rees (1989*b*).

tive, FBPase, operates where they do occur. For example, high activities of PFP are found in non-photosynthetic, non-gluconeogenic tissues in which there is no significant amount of starch (ap Rees, Green & Wilson, 1985). We suggest that net conversion of Fru-1,6-P_2 to Fru-6-P is not the primary role of PFP.

We argue that the case for regarding PFP as a major entry point into cytosolic glycolysis remains not proven. We summarize our major reasons, which we have previously given at length (ap Rees & Dancer, 1987; ap Rees, 1988). First, the maximum catalytic activity of PFP does not correlate with estimates of the rates of glycolysis *in vivo* and does not change in unison with those of the other glycolytic enzymes during development. See for example the data for *Arum* spadix in Table 4, which also demonstrate that PFK does show the behaviour expected of a gly-colytic enzyme. Second, when the rate of glycolysis is varied, either naturally, such as during the onset of thermogenesis in *Arum* spadices, or by exposing tissues to either anoxia or uncouplers, the content of PP_i does not change (Table 5). If PFP regulates entry into glycolysis then we might expect PP_i to change in the opposite direction to flux when glycolysis is varied. We stress that in this type of experiment PP_i is the diagnostic substrate to examine. Arguments based on changes in hexose 6-phos-phates or Fru-2,6-P_2 are not convincing because the former provide substrate for PFK and changes in the latter could reflect changes in the activity of PFP in the direction of PP_i synthesis.

Suggestions that PFP acts to synthesize PP_i in the cytosol require evidence for the presence of PP_i and a role for such a presence. There is appreciable evidence that a wide range of non-photosynthetic and photosynthetic tissues of plants contain PP_i (Edwards et al., 1984; Smyth & Black, 1984; Weiner, Stitt & Heldt, 1987). The latter evidence rests on the assumption that PFP, which is used to assay PP_i in plant extracts, is entirely specific for PP_i and that no other component of the extract will act as a phosphoryl donor to give Fru-1,6-P_2. Support for this assumption is given by the observation that pre-treatment of the extracts with pyrophosphatase abolishes their ability to synthesize Fru-1,6-P_2 via PFP. Further support is given by results from an alternative assay for PP_i that depends upon the enzyme sulphate adenyltransferase.

$$\text{Adenyl sulphate} + PP_i \;\rightleftharpoons\; \text{ATP} + \text{sulphate}$$

Comparison of the two assays gave identical results for the three different tissues examined (Dancer & ap Rees, 1989a). In the same work we showed that the PP_i present in plant extracts was not due to breakdown of phosphoribosyl pyrophosphate during the preparation of the extracts. Indirect evidence from non-photosynthetic tissues (Edwards & ap Rees, 1986; ap Rees, Green & Wilson, 1985) strongly suggested that the PP_i was largely confined to the cytosol. This has now been demonstrated directly for photosynthetic tissues (Weiner, Stitt & Heldt, 1987).

A major function for cytosolic PP_i in plants is suggested by the evidence that plants use sucrose synthase to break down sucrose as follows

$$\text{Sucrose} + \text{UDP} \;\overset{a}{\rightleftharpoons}\; \text{UDP-glucose} + \text{Fru}$$

$$\text{Fru} + \text{UTP} \;\overset{b}{\rightleftharpoons}\; \text{Fru-6-P} + \text{UDP}$$

$$\text{UDP-glucose} + PP_i \;\overset{c}{\rightleftharpoons}\; \text{Glc-1-P} + \text{UTP}$$

$$\text{Fru-1,6-}P_2 + P_i \;\overset{d}{\rightleftharpoons}\; \text{Fru-6-P} + PP_i$$

(a, sucrose synthase; b, fructokinase; c, UDP-glucose pyrophosphorylase; d, PFP)

The essential feature of this pathway, which is confined to the cytosol, is the proposal that there is sufficient PP_i in the cytosol to allow UDP-glucose pyrophosphorylase to convert UDP-glucose, the product of sucrose synthase, to Glc-1-P. This reaction is seen as the crucial link between UDP-glucose and the pool of hexose monophosphates that supplies respiration and the synthesis of starch. The pathway requires one molecule of PP_i per molecule of sucrose broken down. This is because one mole of PP_i will be required for each mole of sucrose broken down by the sucrose synthase pathway. We envisage this PP_i being produced by

PFP at the expense of glycolytically formed Fru-1,6-P_2. This entails substrate cycling between Fru-1,6-P_2 and Fru-6-P. For each molecule of sucrose that enters glycolysis one of the two molecules of Fru-1,6-P_2 produced by PFK will be converted back to Fru-6-P by PFP before finally entering glycolysis via PFK again.

The evidence for this pathway of sucrose breakdown has been given in detail (ap Rees, 1988) and comprises measurements of the maximum catalytic activities of the relevant enzymes, studies of the properties of the enzymes, distribution of ^{14}C after feeding [^{14}C] sucrose, and substrate measurements that strongly suggest that the reactions catalysed by sucrose synthase and UDP-glucose pyrophosphorylase are close to equilibrium *in vivo*. A further possible role for cytosolic PP_i in plants is generation of a proton gradient across the tonoplast via the proton translocating pyrophosphatase (Rea & Saunders, 1987). Thus there is evidence that one of the distinctive features of plant metabolism is the use, to some extent, of PP_i as an energy source in the cytosol.

We must now consider whether production of PP_i is the sole role of PFP in plants. If it is, then we would expect a close relationship between sucrose breakdown, PP_i, and PFP. Our recent experiments (Dancer & ap Rees, 1989c) suggest that both PP_i and PFP are related to sucrose breakdown but not uniquely so. Two particular exceptions strongly suggest an additional role for PFP. First, the enzyme occurs in red algae despite the fact that they lack sucrose. Second, comparison of the developing endosperm of wild type and the *sh 1* shrunken mutant of maize showed that the latter was deficient in sucrose synthase and largely dependent upon invertase for sucrose breakdown. Nonetheless, the mutant did not show any diminution in contents of PP_i or Fru-2,6-P_2, or in the maximum catalytic activities of PFP or UDP-glucose pyrophosphorylase.

The distribution of pyrophosphatase in plant cells provides further evidence that PFP is not solely involved in PP_i production in plants. Alkaline pyrophosphatase is largely confined to the plastid (Gross & ap Rees, 1986; Weiner, Stitt & Heldt, 1987). The tonoplast enzyme is proton-translocating and thus likely to be strictly regulated. This means that the cytosol has no obvious means of removing PP_i formed during biosynthesis of macromolecules and sucrose. The latter syntheses involve the use of pyrophosphorylases whose continued activity requires the removal of PP_i. We suggest that lack of cytosolic pyrophosphatase enables the plant cytosol to maintain a sufficient concentration of PP_i to use it as an energy source, but imposes a need for an alternative means of disposal of PP_i. We suggest that this need is met by PFP, which we envisage as being responsible for the synthesis and breakdown of PP_i, and as acting primarily as a means of maintaining a specific concentration of PP_i in the cytosol. In short we see the enzyme as a 'PP$_i$-stat'. Such a role for PFP is entirely consistent with the observations that the enzyme catalyses a near equilibrium reaction *in vivo*. In this context it seems likely that the main significance of the enzyme's response to Fru-2,6-P$_2$ is that the latter increases the effective maximum catalytic activity and thus allows more efficient regulation of PP$_i$ concentration.

That the concentration of PP$_i$ in the cytosol of plant cells is highly regulated is demonstrated by the fact that it does not change during light–dark transients in leaves (Weiner *et al.*, 1987), during the massive increase in respiration at thermogenesis in *Arum* spadix (ap Rees *et al.*, 1985) or when tissues are made anoxic or are treated with uncouplers (Dancer & ap Rees, 1989b). As Stitt (1989) has pointed out, the properties of PFP are also consistent with it acting as a 'PP$_i$-stat'. The response of the enzyme to its substrates means that flux will not only be affected by the mass action ratio of products to substrates but also by the fact that when PP$_i$ increases then its synthesis will be reduced by inhibition of the back reaction of PFP. Conversely, a fall in PP$_i$ will promote synthesis of PP$_i$.

The use of PFP as a 'PP$_i$-stat' will involve interconversion of Fru-6-P and Fru-1,6-P$_2$. Such interconversions could contribute to net gluconeogenic or glycolytic flux, but in most instances are likely to give rise to substrate cycling. During the breakdown of sucrose net flux will be in the glycolytic direction but PFP will act to produce PP$_i$. During sucrose synthesis there is unlikely to be appreciable cytosolic glycolysis in gluconeogenic and photosynthetic tissues and net flux will be from Fru-1,6-P$_2$ to Fru-6-P with PFP working in the opposite direction to remove PP$_i$. Direct evidence of such substrate cycling has been obtained by Keeling *et al.* (1988) and by Hatzfield & Stitt (1990) and is discussed by Keeling (this volume).

FBPase

Until recently it has been generally assumed that non-photosynthetic cells, other than those involved in the conversion of fat to sugar, lack FBPase. However, renewed interest in starch metabolism by non-photosynthetic cells has led to reports of the enzyme in suspension cultures of soybean (Macdonald & ap Rees, 1983), cauliflower florets (Journet & Douce, 1985), and endosperm of maize (Echeverria *et al.*, 1988) and wheat (Sangwan & Singh, 1988). As argued earlier, the issue is central to what crosses the plastid membrane in respect of starch metabolism in non-photosynthetic cells. Here we use data obtained from a study of wheat endosperm to urge caution before accepting the view that such non-photosynthetic tissues contain FBPase.

Table 6. *Assay of unfractionated extracts of wheat leaf and endosperm for plastidic and cytosolic FBPase*

Tissue	Enzyme	Fru-6-P formation (nmol min^{-1} g^{-1} fresh wt)	
		Minus Fru-2,6-P$_2$	Plus Fru-2,6-P$_2$ (20 μM)
Leaf	Plastidic	273±16	220±25
	Cytosolic	79±24	None found
Endosperm	Plastidic	95±12	125±24
	Cytosolic	13± 4	83±27

Data from Entwistle & ap Rees (1988): values are means ± SE from at least 3 extracts.

Detection of whether a tissue contains a particular enzyme is generally achieved by careful assay of an unfractionated extract: any extensive fractionation creates the risk of losing what you are looking for. If this is done for FBPase, the presence of PFP will result in the conversion of Fru-1,6-P$_2$ to Fru-6-P and apparent FBPase activity will be seen. This is because the extract and, or, the Fru-1,6-P$_2$ used in the assay will almost certainly contain or give rise to sufficient P$_i$ to allow PFP to function. The requirement for Fru-2,6-P$_2$ will either be met by contamination of Fru-1,6-P$_2$ by Fru-2,6-P$_2$ or by the relatively high concentration of Fru-1,6-P$_2$ used in the assay. However we should be able to distinguish between FBPase and PFP on the basis of their response to Fru-2,6-P$_2$. FBPase from the cytosol is allosterically inhibited by Fru-2,6-P$_2$, that from the chloroplast is inhibited competitively. PFP is stimulated by Fru-2,6-P$_2$. Accordingly, we assayed extracts of wheat leaves and wheat endosperm for plastidic and cytosolic FBPase in the presence and absence of Fru-2,6-P$_2$.

FBPase was detected in extracts of the wheat leaves (Table 6): the cytosolic form was severely inhibited by Fru-2,6-P$_2$, the plastidic form slightly so. These results show that it is possible to detect FBPase despite the presence of PFP. Assay of endosperm extracts led to Fru-6-P formation, but this was stimulated by Fru-2,6-P$_2$ not inhibited, suggesting that the activity represented PFP. This suggestion is confirmed by the data in Table 7, which show that antibody to PFP had no effect on FBPase activity detected in leaf extracts but abolished that found in endosperm extracts. Pre-immune serum was without effect.

Table 7. *Effect of antibody to PFP on conversion of Fru-1,6-P$_2$ to Fru-6-P by extracts of wheat endosperm and leaves*

| | Fru-6-P formation (nmol min^{-1} g^{-1} fresh wt) | | | |
| | Pre-immune serum | | Antibody | |
Tissue	Minus Fru-2,6-P$_2$	Plus Fru-2,6-P$_2$ (20 µM)	Minus Fru-2,6-P$_2$	Plus Fru-2,6-P$_2$ (20 µM)
Endosperm	373± 6	521± 9	None found	
Leaf	316, 478	282, 452	394, 485	332, 452

Data from Entwistle & ap Rees (1988).

In a complementary approach to the same question we used antibodies to plastidic FBPase from spinach leaves. We prepared unfractionated extracts of wheat leaves, endosperm, and of a mixture of leaves and endosperm. Proteins were separated by polyacrylamide gel electrophoresis, transferred to nitrocellulose paper and analysed for FBPase by immunodecoration with the FBPase antibody and ^{125}I-labelled Protein A (Fig. 1). The antibody recognized protein in the wheat leaves and did so even when the leaves were homogenized with endosperm. No protein was recognized in the endosperm extracts, strongly suggesting that the latter lack FBPase. Enzyme assay of purified amyloplasts also failed to reveal any FBPase.

If our contention that wheat endosperm lacks FBPase is right then carbon for starch synthesis does not enter the wheat amyloplast as triose phosphate. Two independent pieces of evidence suggest that this is so. First, the extensive labelling data of Keeling *et al.*, 1988 (and Keeling, this volume) show that the conversion of glucose to starch by wheat endosperm involves relatively little randomization between carbons 1 and 6: extensive randomization via cytosolic and plastidic triosephosphate isomerase would be expected if the triose pathway operated. Further, the extent of such randomization that did occur is almost identical to that found in the hexosyl groups of sucrose. This has already been attributed to the action of PFP, and strongly suggests that sucrose and starch derived their hexosyl groups from the same pool of cytosolic hexose monophosphates.

Final support for the view that wheat endosperm lacks an FBPase is provided by studies with isolated amyloplasts (Tyson & ap Rees, 1988). When these plastids were incubated in labelled substrates they incorporated Glc-1-P into starch. This incorporation was dependent upon

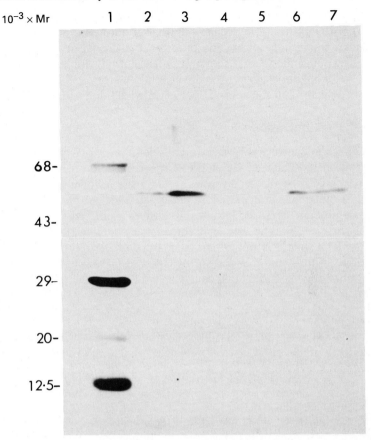

Fig. 1. Identification of plastidic FBPase in wheat extracts by immunoblotting. Lane 1, standard proteins; lane 2, 50 μg of wheat leaf protein; lane 3, 100 μg of wheat leaf protein; lane 4, 50 μg of endosperm protein; lane 5, 100 μg of endosperm protein; lane 6, 50 μg of a mixture of leaf and endosperm protein; lane 7, 100 μg of a mixture of leaf and endosperm protein.

plastid integrity. No such incorporation was found when labelled triose phosphates were presented to the plastids. Thus our inability to detect FBPase in wheat amyloplasts correlates with independent evidence that it is a C-6 compound, not a C-3 compound, that crosses the amyloplast envelope to provide the carbon for starch synthesis. In view of our results we suggest that claims that non-photosynthetic, non-gluconeogenic tissues contain FBPase must be accompanied by proof that the activity is not

PFP. Such evidence has not been published in respect of the claims that we listed earlier. Our recent work has confirmed that wheat endosperm lacks significant activity of plastidic FBPase (Entwistle & ap Rees, 1990a), and shown that the same is true for developing tubers of potato, developing endosperm of maize, cauliflower florets, suspension cultures of soybean, and the roots of peas (Entwistle & ap Rees, 1990b).

Conclusion

Our main conclusion is that a great deal more work is needed before the relationship between C-6 and C-3 sugar phosphates is established. Such work might be directed towards testing the following three hypotheses. First, net conversion of hexose phosphate to triose phosphate resulting in commitment of carbon to respiration is mediated in the cytosol and plastid via PFK. Second, net conversion of triose phosphate to hexose monophosphate is not a normal feature of non-photosynthetic cells of plants. Third, interconversion of C-6 and C-3 sugar phosphates occurs continually via PFP acting as a 'PP$_i$-stat'.

References

ap Rees, T. (1977). Conservation of carbohydrate by the non-photosynthetic cells of higher plants. *Symp. Soc. Exp. Biol.* **31**, 7–32.

ap Rees, T. (1985). The organization of glycolysis and the oxidative pentose phosphate pathway in plants. In *Encyclopedia of Plant Physiology*, vol. 18, ed. R. Douce & D. A. Day, pp. 391–417. Berlin: Springer-Verlag.

ap Rees, T. (1988). Hexose phosphate metabolism by nonphotosynthetic tissues of higher plants. In *The Biochemistry of Plants*, vol. 14, ed. J. Priess, pp. 1–33. New York: Academic Press.

ap Rees, T. & Dancer, J. E. (1987). Fructose-2,6-bisphosphate and plant respiration. In *Plant Mitochondria: Structural, Functional and Physiological Aspects*, ed. A. L. Moore & R. B. Beechey, pp. 341–50. New York: Plenum Press.

ap Rees, T., Green, J. H. & Wilson, P. M. (1985). Pyrophosphate fructose 6-phosphate 1-phosphotransferase and glycolysis in non-photosynthetic tissues of higher plants. *Biochem. J.* **227**, 299–304.

Copeland, L. &. Turner, J. F. (1987). The regulation of glycolysis and the pentose phosphate pathway. In *The Biochemistry of Plants*, vol. 11, ed. D. D. Davies, pp. 107–28. New York: Academic Press.

Dancer, J. E. & ap Rees, T. (1989a). Phosphoribosyl pyrophosphate and the measurement of inorganic pyrophosphate in plant tissues. *Planta* **177**, 261–4.

Dancer, J. E. & ap Rees, T. (1989b). Effects of 2,4-dinitrophenol and anoxia on the inorganic pyrophosphate content of the spadix of *Arum* maculatum and the root apices of *Pisum sativum. Planta*, **178**, 421–4.

Dancer, J. E. & ap Rees, T. (1989c). Relationship between pyrophosphate fructose 6-phosphate 1-phosphotransferase, sucrose breakdown, and respiration. *J. Plant Physiol.* **135**, 197–206.

Dietz, K.-J. & Heber, U. (1984). Rate-limiting factors in leaf photosynthesis. *Biochim. Biophys. Acta* **767**, 432–43.

Dixon, W. L. & ap Rees, T. (1980). Identification of the regulatory steps of glycolysis in potato tubers. *Phytochemistry* **19**, 1297–301.

Echeverria, E., Boyer, C. D., Thomas, P. A., Liu, K.-C. & Shannon, J. C. (1988). Enzymic activities associated with maize kernel amyloplasts. *Plant Physiol.* **86**, 782–92.

Edwards, J. & ap Rees, T. (1986). Metabolism of UDP-glucose by developing embryos of round and wrinkled varieties of *Pisum sativum. Phytochemistry* **25**, 2033–9.

Edwards, J., ap Rees, T., Wilson, P. M. & Morrell, S. (1984). Measurement of inorganic pyrophosphate in tissues of *Pisum sativum* L. *Planta* **162**, 188–91.

Entwistle, G. & ap Rees, T. (1988). Enzymic capacities of amyloplasts from wheat (*Triticum aestivum*) endosperm. *Biochem. J.* **295**, 391–6.

Entwistle, G. & ap Rees, T. (1990a). Failure to corroborate claims that the developing endosperm of wheat (*Triticum aestivum*) contains significant activities of fructose-1,6-bisphosphatase. *Physiologia Pl.* **79**, 635–40.

Entwistle, G. & ap Rees, T. (1990b). Lack of fructose-1,6-bisphosphatase in a range of higher plants that store starch. *Biochem. J.***271**, 467–472.

Gross, P. & ap Rees, T. (1986). Alkaline inorganic pyrophosphatase and starch synthesis in amyloplasts. *Planta* **167**, 140–5.

Hatzfeld, W.-D. & Stitt, M. (1990). A study of the rate of recycling of triose phosphates in heterotrophic *Chenopodium rubrum* cells, potato tubers, and maize endosperm. *Planta* **180**, 198-204.

Ho, L. C. (1988). Metabolism and compartmentation of imported sugars in sink organs in relation to sink strengths. *Ann. Rev. Plant Physiol.* **39**, 355–73.

Journet, E.-P. & Douce, R. (1985). Enzymic capacities of purified cauliflower bud plastids for lipid synthesis and carbohydrate metabolism. *Plant Physiol.* **79**, 458–67.

Keeling, P. L., Wood, J. R., Tyson, R. H. & Bridges, I. G. (1988). Starch biosynthesis in developing wheat grain. *Plant Physiol.* **87**, 311–19.

Kombrink, E., Kruger, N. J. & Beevers, H. (1984). Kinetic properties of pyrophosphate: fructose-6-phosphate phosphotransferase from germinating castor bean endosperm. *Plant Physiol.* **74**, 395–401.

Kruger, N. J. & Beevers, H. (1984). Effect of fructose-2,6-bisphosphate on the kinetic properties of cytoplasmic fructose-1,6-bisphosphatase from germinating castor bean endosperm. *Plant Physiol.* **76**, 49–54.

Kruger, N. J. & Hammond, J. B. W. (1988). Molecular comparison of pyrophosphate and ATP-dependent fructose 6-phosphate 1-phosphotransferases from potato tuber. *Plant Physiol.* **86**, 645–8.

Kruger, N. J., Hammond, J. B. W. & Burrell, M. M. (1988). Molecular characterization of four forms of phosphofructokinase purified from potato tuber. *Arch. Biochem. Biophys.* **267**, 690–700.

Leegood, R. C. & ap Rees, T. (1978). Identification of the regulatory steps in gluconeogenesis in cotyledons of *Cucurbita pepo*. *Biochim. Biophys. Acta* **542**, 1–11.

Macdonald, F. D. & ap Rees, T. (1983). Enzymic properties of amyloplasts from suspension cultures of soybean. *Biochim. Biophys. Acta* **755**, 81–9.

Newsholme, E. A. & Start, C. (1973). *Regulation in Metabolism*. London: Wiley.

Rea, P. A. & Saunders, D. (1987). Tonoplast energization: two H^+ pumps, one membrane. *Physiologia Pl.* **71**, 131–41.

Sangwan, R. S. & Singh, R. (1988). Two forms of fructose 1,6-bisphosphatase from immature wheat endosperm. *Physiologia. Pl.* **73**, 21–6.

Smyth, D. A. & Black, C. C. (1984). Measurement of pyrophosphate content in plant tissues. *Plant Physiol.* **75**, 862–4.

Stitt, M. (1989). Product inhibition of potato tuber pyrophosphate fructose-6-phosphate phosphotransferase by phosphate and pyrophosphate. *Plant Physiol.* **89**, 628–33.

Stitt, M., Lilley, R. McC. & Heldt, H. W. (1984). Adenine nucleotide levels in the cytosol, chloroplasts and mitochondria of wheat leaf protoplasts. *Plant Physiol.* **70**, 971–7,

Stitt, M., Wirtz, W. & Heldt, H. W. (1980). Metabolite levels in the chloroplast and extra-chloroplast compartments of spinach protoplasts. *Biochim. Biophys. Acta* **593**, 85–102.

Stitt, M., Huber, S. & Kerr, P. (1987). Control of photosynthetic sucrose formation. In *The Biochemistry of Plants*, vol. 10, ed. M. D. Hatch & N. K. Boardman, pp. 327–409. New York: Academic Press.

Tyson, R. H. & ap Rees, T. (1988). Starch synthesis by isolated amyloplasts from wheat endosperm. *Planta* **175**, 33–8.

Weiner, H., Stitt, M. & Heldt, H. W. (1987). Subcellular compartmentation of pyrophosphate and alkaline pyrophosphatase in leaves. *Biochim. Biophys. Acta* **893**, 13–21.

Wirtz, W., Stitt, M. & Heldt, H. W. (1980). Enzyme determination of metabolites in the subcellular compartments of spinach protoplasts. *Plant Physiol.* **66**, 187–95.

PETER L. KEELING

The pathway and compartmentation of starch synthesis in developing wheat grain

Introduction

Starch is a major component of the 'average' dietary intake of man and animals. Calculated in calories (Borgstrom, 1973), about four-fifths of the world's food is provided by three grain crops (maize, wheat and rice) and three tuber crops (potato, yam and cassava). On a dry-weight basis starch is by far the major component of the edible portions of these crops, providing between 60 to 90% of the dry weight. As well as its uses in nutrition, starch is also an important component in manufacturing a wide range of industrial products such as paper, textiles and building materials. Furthermore, chemically modified starch and starch derivatives are used widely throughout industry. World-wide, maize represents the major commercial source of starch, whereas wheat starch is of only minor significance in the starch industry.

Despite the real importance of starch in our diets and as a renewable resource for industry, there remains much to be learnt about how starch is stored and metabolized in plants. This paper will concentrate on just one area of ignorance that has recently been significantly eroded by progress in research on the pathway of starch synthesis in developing wheat grain. Where relevant, recent publications about other starch storing crops, mainly maize, will also be covered.

Starch deposition in plants

Starch plays a central role in metabolism in most higher plants since it serves as the major food reserve. It is deposited as water-insoluble granules as a mixture of amylose and amylopectin. In chloroplasts, starch acts as a temporary store during photosynthesis: in the dark the starch is remobilized. Longer-term storage takes place in the reserve organs during one phase of the growth of the plant: the starch will be used at another time for germination. In plant storage tissues such as in the developing

111

cereal grain, the final stages of starch biosynthesis are confined to a separate metabolic compartment within the cytosol. This intracellular organelle is termed an amyloplast and consists of a starch granule surrounded by plastid stroma which is enclosed by a double membrane.

Based on *in vitro* assays of various enzymes present in whole-tissue extracts, a biosynthetic pathway has been proposed (Turner, 1969) which became generally accepted in the literature (Fig. 1). However, because

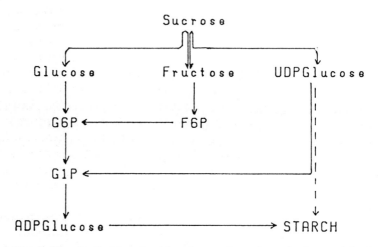

Fig. 1. The classical 'textbook' pathway of starch synthesis as originally proposed by Turner (1969).

this classical 'textbook' pathway does not take into account the intracellular compartmentation of starch synthesis within the amyloplast and the selectively permeable nature of plastid membranes, we cannot confidently say how translocated sucrose is converted into starch. Virtually all of the published research on the selectively permeable nature of plastid membranes has been with chloroplasts (Heber, 1974; Heber & Heldt, 1981), although the transport properties of chromoplasts isolated from daffodil flowers (Liedvogel & Kleinig, 1980) and of oily plastids isolated from developing endosperm of castor oil seeds, have been reported (Miernyk & Dennis, 1983). All these plastids appear to share common properties in their ability to transport triose phosphates such as 3-phosphoglycerate and dihydroxyacetone phosphate and to some extent glucose. Since amyloplasts and chloroplasts both develop from proplastids and under certain conditions amyloplasts develop into chloroplasts (Muhlethaler, 1971), and vice-versa (Badenhuizen, 1969), it has been suggested that triose phosphates are transported into the amyloplast in a manner analogous to that found in chloroplasts (Jenner, 1976). Thus an alternative metabolic pathway has been proposed (Boyer, 1985; Shannon & Garwood, 1985) for starch formation in storage tissues which is similar to that of starch and sucrose formation in leaves, and involves triose phosphates.

The last steps in the pathway

The last steps in the pathway are confined to the amyloplast, where starch synthesis is mediated by the specific transfer to starch of the glucosyl units of ADP-glucose by the enzyme starch synthase in combination with branching enzyme. Furthermore, since ADP-glucose pyrophosphorylase is confined to the amyloplast (MacDonald & ap Rees, 1983; Journet & Douce, 1985; Entwistle & ap Rees, 1988) the main source of ADPG must involve its synthesis from glucose 1-phosphate. The early steps of this pathway must, therefore, fit together to provide this sugar-phosphate. The maize mutants, waxy, dull, amylose extender and shrunken-2 or brittle-2 provide evidence that bound and soluble starch synthases, branching enzymes and ADP-glucose pyrophosphorylase are important steps in the pathway of starch synthesis in maize endosperm (Shannon & Garwood, 1985). Similarly, branching enzyme must be a key step in starch synthesis in pea cotyledons because the wrinkled mutant of pea has been shown to affect branching enzyme activity (Matters & Boyer, 1982; Smith, 1988). These mutant phenotypes of maize or pea are characterized by having significant reductions in starch content or in amylose and amylopectin content (Shannon & Garwood, 1985).

The early pathway

Evidence from wheat grain suggests that the first step in the pathway must involve sucrose synthase, because the alternative enzyme (-invertase-) was found to be absent from wheat endosperm tissue (Chevalier & Lingle, 1983). If this is correct, the second steps in the pathway must involve UDP-glucose pyrophosphorylase and hexokinase, producing glucose 1-phosphate and fructose 6-phosphate, respectively. This helps to provide other evidence for the involvement of sucrose synthase in the early pathway. Thus, the enzyme hexokinase, which catalyses the conversion of hexose to hexose 6-phosphate, has been shown to have a much higher affinity for fructose than for glucose when isolated from wheat germ (Higgins & Easterby, 1976). Furthermore the enzyme UDP-glucose pyrophosphorylase is extremely active in wheat endosperm tissue (Turner, 1969; Kumar & Singh, 1984) and may well be able to catalyse the pyrophosphate-dependent formation of glucose 1-phosphate from UDP-glucose because pyrophosphate levels are appreciable in this tissue (Dancer & ap Rees, 1989).

The first step in the pathway of starch synthesis in developing maize grain has been attributed to invertase as well as sucrose synthase. Enzyme distribution studies (Shannon & Dougherty, 1972; Doehlert & Felker, 1987) showed invertase activity to be highest in the basal endosperm, whereas sucrose synthase is located mainly in the upper endosperm tissue. Consistent with this is the finding that glucose and fructose accumulate in the basal regions of the maize kernel (Shannon, 1972; Hanft & Jones, 1986). The endosperm mutant shrunken-1 provides evidence that sucrose synthase must also play a significant role in the pathway of starch synthesis in maize (Shannon & Garwood, 1985). Thus the first steps in the pathway of starch synthesis in maize must involve invertase in combination with sucrose synthase. The second steps in the pathway of starch synthesis in developing maize grain must involve UDP-glucose pyrophosphorylase as well as glucokinase and fructokinase, producing glucose 1-phosphate, glucose 6-phosphate and fructose 6-phosphate respectively. All three enzymes have been shown to be present in developing maize grain endosperm tissue (Doehlert, Kuo & Felker, 1988). Other evidence is emerging that an NADH-dependent ketose reductase is present in maize endosperm (Doehlert, 1987). However, the precise function of this enzyme in the pathway of starch synthesis needs investigating further.

In summary, these data from developing wheat and maize grain show that the glucosyl moiety of sucrose is converted, first to either UDP-glucose or glucose and then to glucose 1-phosphate or glucose 6-phosphate, whereas the fructosyl moiety is converted, first to fructose and then to fructose 6-phosphate.

Plastid enzyme compartmentation and transport properties

There have been many reports of isolation of amyloplasts from storage organs. However, not all of these have provided adequate biochemical proof that the amyloplast preparation was both intact and also free from contamination with other cellular compartments. Where such data was not provided there is cause for concern over the interpretation of the results. Therefore, the present chapter will focus only on those reports which provide such proof.

Evidence in support of the concept that triose phosphates are involved in the biosynthetic pathway has been provided by several groups of workers (Liu & Shannon, 1981; MacDonald & ap Rees, 1983; Journet &

Douce, 1985). Using a non-aqueous amyloplast isolation procedure Liu & Shannon (1981) showed that the soluble extracts of a starch granule preparation contained varying amounts of neutral sugars, inorganic phosphate, hexose and triose phosphates, organic acids and amino acids, nucleotides and sugar nucleotides. MacDonald & ap Rees (1983) and Journet & Douce (1985), used an aqueous isolation procedure to prepare an intact amyloplast fraction that contained the plastid stromal enzymes with little contamination by cytosol or by other organelles. These stromal enzymes included all those needed to convert triose phosphates to starch. On the basis of these data it has been suggested that the entry of carbon for starch synthesis could take place via a phosphate translocator.

However, more recently, techniques have been developed using developing wheat grain endosperm tissue for the isolation of intact amyloplasts which retain their transport properties (Tyson & ap Rees, 1988). These authors have demonstrated that glucose 1-phosphate is taken up and incorporated into starch, whereas triose phosphates are not. Thus, glucose 1-phosphate may be the most likely compound that crosses the amyloplast membrane to support starch synthesis. Further work with the isolated amyloplasts demonstrated that wheat endosperm lacks significant plastidic fructose-1,6-bisphosphatase (Entwistle & ap Rees, 1988), indicating that even though amyloplasts may possess a phosphate translocator (Alban, Joyard & Douce, 1988), uptake of triose phosphates cannot be the source of carbon for starch synthesis. However, these data and conclusions are at variance with the recent work of Echeverria *et al.* (1988) on amyloplasts isolated from developing maize kernels. These authors found similar enzyme-distributions to that previously reported by other workers, but they also found significant fructose-1,6-bisphosphatase activity in their amyloplast fraction. Furthermore, using ^{14}C labelled substrates they obtained significant starch synthesis from dihydroxyacetone phosphate with intact amyloplasts. Since these studies were done in different plant species, it is possible that amyloplasts differ in their transport properties. However, there are some concerns over the data obtained from developing maize grain. Firstly, the uptake of triose phosphates was not abolished by lysing the amyloplast membrane. Secondly, ^{14}C incorporation into the starch in the methanol-insoluble fraction was not proven by hydrolysing the starch with enzymes and showing that the label released was associated exclusively with glucose. Thirdly, uptake was not shown to be time-dependent. Because of these concerns it is difficult to draw an unequivocal conclusion from the data. Whether maize endosperm amyloplasts are truly different from wheat endosperm amyloplasts requires further investigation.

With the currently available evidence from studies of the transport properties of amyloplasts isolated from developing wheat grain, glucose 1-phosphate is the most likely compound that crosses the amyloplast membrane to support starch synthesis. These data would, therefore, suggest that the most likely pathway of starch synthesis in non-photosynthetic tissues is similar to that originally proposed (Turner, 1969; Fig. 1), whereas the alternative metabolic pathway involving triose-phosphates (Boyer, 1985; Shannon & Garwood, 1985) now seems unlikely. However, certain aspects of the classical textbook pathway of starch synthesis must be incorrect. For example, the direct transfer of glucosyl units from UDP-glucose to starch is unlikely, since UDP-glucose pyrophosphorylase is confined to the cytoplasm, whereas ADP-glucose pyrophosphorylase is confined to the amyloplast (MacDonald & ap Rees, 1983; Journet & Douce, 1985). Furthermore, the classical pathway omits the step involved in transferring carbon from the cytoplasm to the amyloplast compartments. Another missing link in the classical pathway is the involvement of pyrophosphate in the pyrophosphorylase reactions involving UDP-glucose and ADP-glucose. The recent finding (Gross & ap Rees, 1986) that alkaline inorganic pyrophosphatase is absent from the cytoplasm and confined to the amyloplast is important as it provides a means to minimize pyrophosphate levels in the amyloplast. This will favour the formation of ADP-glucose in the amyloplast and favour the formation of glucose 1-phosphate in the cytoplasm; reactions that would otherwise be in equilibrium. A means to generate pyrophosphate in the cytoplasm has been speculated to involve pyrophosphate-dependent phosphofructokinase (ap Rees, Green & Wilson, 1985; Huber & Akazawa, 1986), although other workers (Doehlert, Kuo & Felker, 1988) favour an exchange of plastidic pyrophosphate with cytoplasmic ATP via an adenylate/PP_i exchanger such as that found in pea chloroplasts (Robinson & Wiskich, 1977). These hypotheses will be discussed again later in the light of the data presented in the next section.

A direct approach to evaluating the involvement of triose phosphates

The reasons for the proposal that triose phosphates may be involved in the metabolic pathway of starch synthesis have already been presented. However, based on the evidence for the absence from isolated amyloplasts of fructose-1, 6-bisphosphatase and selective uptake of glucose 1-phosphate it seems likely that a hexosyl sugar may be transferred instead. Using NMR spectroscopy and ^{13}C-labelled sugars a more direct examination of the involvement of triose phosphates in the metabolic pathway has been shown to be possible (Keeling et al., 1988). This technique provides an extremely attractive alternative to the use of radio-isotopes as the data are so readily obtained with little need for laborious extraction and processing procedures. The only drawback is the low-sensitivity of ^{13}C-NMR spectroscopy compared with using radiotracers. With this technique it was possible to quantify the extent of redistribution of ^{13}C-label between carbons 1 and 6 of glucose incorporated into starch in intact endosperm tissue. This permitted a direct estimate of the degree of involvement of triose-phosphates, because the enzyme triose phosphate isomerase will redistribute the label between the C-1 and C-6 carbons of any hexoses incorporated into starch (Katz, Landau & Bartsch, 1966). In the work of Keeling *et al.* (1988) it was assumed that the triose phosphates, glyceraldehyde-3 phosphate and dihydroxyacetone phosphate, were in equilibrium within the cell, and that any resynthesis of fructose-1, 6-bisphosphate from triose-phosphate occurred with random triose units. The activity of triose-phosphate isomerase in wheat endosperm tissue has been shown to be many fold higher than other enzymes involved in carbohydrate metabolism (Entwistle & ap Rees, 1988). Thus, since the activity of this enzyme is many fold in excess of the rate of starch synthesis, we suggested that in developing wheat endosperm there was likely to be full isotopic equilibrium between the triose phosphates. Therefore, if triose-phosphates are involved in the metabolic pathway of starch synthesis, starch synthesized from glucose labelled specifically in the C-1 or C-6 positions should be found to have a high degree of redistribution of label between the C-1 and C-6 carbons. Conversely, if the metabolic pathway of starch biosynthesis involves a series of hexose inter-conversions, without the direct involvement of triose phosphates, starch synthesized from such asymmetrically-labelled glucose should be found to have a highly conserved degree of distribution of label.

When wheat endosperm was supplied *in vitro* and *in vivo* with [1-^{13}C] labelled glucose or fructose there was a consistent pattern of partial label redistribution in C-1 and C-6 of glucose incorporated into starch and into the glucosyl and fructosyl moieties of sucrose (Table 1). This amounted to about 15% of the label incorporated, with no detectable label incorpora-tion into carbons 2 to 5. The degree of partial redistribution was remark-ably constant despite various changes in the extracellular environment which significantly affected either the amount of radioactivity incorporated or the rate of starch synthesis achieved. Further experi-ments involved incubating wheat endosperm with [6-^{13}C] glucose and comparing this with incubations with [1-^{13}C] glucose. In this case there was a similar degree of label redistribution in the glucosyl and fructosyl moieties of sucrose but label redistribution in starch was about half that found in sucrose (Table 2). This difference was attributed to a preferen-

Table 1. In vitro *and* in vivo *experiments with [1-^{13}C] glucose: percentage redistribution ratios in sucrose and starch extracted from wheat endosperm*

| | | ^{13}C redistribution from C1 to C6 (%) | |
		in vitro	*in vivo*
Sucrose	Glucosyl:Fructosyl	57:43	57:43
Sucrose	Glucosyl C1:C6	15·9	21·8
Sucrose	Fructocosyl C1:C6	17·1	29·1
Sucrose	Average C1:C6	16·5	25·5
Starch	(Glucose) C1:C6	19·3	27·8 (±1·1)

In vitro experiment: isolated endosperm tissue from 400 grain at 21 days post anthesis was incubated *in vitro* for 5 h at 25° C with [1–^{13}C] glucose. *In vivo* experiment: plants at 21 days post anthesis were supplied with [1–^{13}C] glucose via a nick in the stem. After 1 and 3 days after supplying the label the endosperm tissue was dissected from 40 grain. The ^{13}C-starch and ^{13}C-sucrose (produced by metabolism in the endosperm tissue) was isolated, after homogenizing the tissue in the ice-cold 1 M perchloric acid. The starch was partially hydrolysed to glucose using amyloglucosidase and the sucrose was purified by HPLC. The distribution of ^{13}C- in glucose hydrolysed from starch or purified sucrose was measured by NMR spectrometry.
Source: (Keeling *et al.*, 1988).

tial loss of label from the C-1 of hexose incorporated into starch possibly via the oxidative pentose phosphate pathway. Thus, the redistribution ratio in starch for [1-^{13}C] glucose (18%) is an overestimate, whereas the ratio for [6-^{13}C] glucose (8%) is an underestimate – the true value must lie somewhere between the two.

Collectively, these data clearly show that the label redistribution pattern seen in starch is the same as that seen in sucrose. Thus, in terms of the C-1 to C-6 label redistribution patterns, the hexose sugars available for sucrose biosynthesis are in equilibrium with the hexose sugars available for starch biosynthesis. Since it seems most unlikely that sucrose is present in the amyloplast (MacDonald & ap Rees, 1983; Entwistle & ap Rees, 1988; Echeverria *et al.*, 1988) it appears that the cytoplasmic sugars used for sucrose biosynthesis are in equilibrium with the amyloplastic sugars used for starch biosynthesis. These findings seriously weaken the case for the direct involvement of triose phosphates in the metabolic pathway of starch biosynthesis. Furthermore, such data are clearly con-

Table 2. In vitro *experiments with [1-^{13}C] and [6-^{13}C] glucose: percentage redistribution ratios in sucrose and starch extracted from wheat endosperm*

		^{13}C redistribution from C1 to C6 (%)	
		[1-^{13}C]	[6-^{13}C]
Sucrose	Glucosyl:Fructosyl	58:42	55:45
Sucrose	Glucosyl C1:C6	16·0	15·2
Sucrose	Fructocosyl C1:C6	19·7	14·7
Sucrose	Average C1:C6	17·9	15·0
Starch	(Glucose) C1:C6	17·7 (±1·0)	8·1 (±0·1)

Isolated endosperm tissue from 100 grain at 21 days post anthesis was incubated *in vitro* for 5 h at 25° C with [1–^{13}C] or [6–^{13}C] glucose. The ^{13}C-starch and ^{13}C-sucrose (produced by metabolism in the endosperm tissue) was isolated, after homogenizing the tissue in the ice-cold 1 M perchloric acid. The starch was partially hydrolysed to glucose using amyloglucosidase and the sucrose was purified by HPLC. The distribution of ^{13}C- in glucose hydrolysed from starch or purified sucrose was measured by NMR spectrometry. Results of sucrose measurements are expressed as the mean of one determination from 100 grain. *Source:* (Keeling *et al.*, 1988).

sistent with the findings that a hexose phosphate is transported into isolated amyloplasts (Tyson & ap Rees, 1988) and that such amyloplasts lack significant fructose-1,6-bisphosphatase activity (Entwistle & ap Rees, 1988). It is now possible to argue that the sugar transported across the amyloplast membrane must be a hexose and that glucose 1-phosphate is the most likely candidate, at least in developing wheat grain.

Conclusions

Based on these data, it is possible to propose a working hypothesis for the metabolic pathway of starch biosynthesis in developing wheat grain, which is shown in Fig. 2. Fructose and UDP-glucose are the first products of sucrose hydrolysis by the enzyme sucrose synthase. Glucose 1-phosphate is formed from UDP-glucose by UDP-glucose pyrophosphorylase and fructose 6-phosphate is formed from fructose by hexokinase. Two further enzymes, phosphoglucomutase and glucose phosphate isomerase

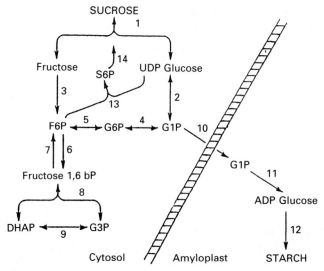

Enzymes

1 Sucrose synthase
2 UDP-glucose pyrophosphorylase
3 Hexokinase
4 Phosphoglucomutase
5 Hexose phosphate isomerase
6 ATP-dependent phosphofructokinase
7 Fructose-1,6-bisphosphatase
and PPi-dependent
phosphofructokinase

8 Aldolase
9 Triose phosphate isomerase
10 Hexose translocator
11 ADP-glucose pyrophosphorylase
12 Starch synthase
13 Sucrose phosphate synthase
14 Sucrose phosphatase

Fig. 2. The proposed metabolic pathway of starch synthesis in developing wheat grain.

are then responsible for the interconversion of the hexose phosphates. Resynthesis of sucrose from the cytoplasmic hexoses is then possible via the enzymes sucrose synthase, sucrose phosphate synthase and sucrose phosphatase. From our findings, a seemingly constant proportion (between 30–40%) of the hexose sugars then enter a cytoplasmic and rapid pathway cycle between the hexose phosphates and triose phosphates where label redistribution occurs between the C-1 and C-6 carbons. Interestingly, other workers have measured the proportion of label redistributed between C-1 and C-6 carbons and this is remarkably constant in several plant species and plant parts (Table 3). To my knowledge this is the first evidence for a pathway cycle in plant cells, but the finding is analogous to that proposed in animal tissues (Clark *et al.*, 1973; Rognstad & Katz, 1980). It is possible that this pathway cycle serves some

Table 3. *Redistribution of label between C1 and C6 carbons incorporated into various plant species and plant parts*

	Redistribution ratio in C1 to C6 carbons (%)
Tobacco leaf starch and sucrose (MacLachlan & Porter, 1959)	15
Wheat straw cellulose (Brown & Neish, 1954)	14·5
Pea cotyledon glucose-6-phosphate (Wager, 1983)	30
Wheat seedling sucrose and cellulose (Edelman *et al.*, 1955; Neish, 1955)	15·3 and 12·3
Cotton Boll Cellulose (Shafizadey & Wolfram, 1955)	6·6
Strawberry and Boysenberry fruit pectin (Seegmiler *et al.*, 1955; 1956)	14

Source: (Keeling *et al.*, 1988).

important function within the cell, perhaps by providing PP_i, from PP_i-dependent phosphofructokinase at the expense of ATP, from ATP-dependent phosphofructokinase (ap Rees *et al.*, 1985; Huber & Akazawa, 1986).

Finally, these data show that glucose 1-phosphate is the most likely sugar being transported into the amyloplast where it is incorporated into starch, via the amyloplastic enzymes ADP-glucose pyrophosphorylase and starch synthase.

Acknowledgements

The research reported here was conducted in the ICI research laboratories at Runcorn, Cheshire, UK (formerly ICI Corporate Bioscience Laboratory). The author thanks Mr David Holt, Miss Philippa Bacon, Mr Philip James and Miss Sheila Baird for their excellent technical assistance. I am also grateful to Dr Ian Bridges, Dr Huw Tyson and Dr Tom ap Rees for their constructive criticism during the course of this work. Finally, I am grateful to Mr John Woods (now deceased) and Dr

Lee Griffiths and other members of the NMR Spectroscopy Group at Runcorn for their valuable expert knowledge.

References

Alban, C., Joyard, J. & Douce, R. (1988). Preparation and characterisation of envelope membranes from nongreen plastids. *Plant Physiol.* **88**, 709–17.

ap Rees, T., Green, J. H. & Wilson, P. M. (1985) Pyrophosphate: fructose 16-phosphate 1-phosphotransferase and glycolysis in nonphotosynthetic tissues of higher plants. *Biochem. J.* **227**, 229–304.

Badenhuizen, N. P. (1969). *The Biogenesis of Starch Granules in Higher Plants.* New York: Appleton–Century Crofts.

Borgstrom, G. (1973). *World Food Resources.* Aylesbury: International Textbook Co. Ltd.

Boyer, C. D. (1985). Synthesis and breakdown of starch. In *Biochemical Basis of Plant Breeding*, vol. I. *Carbon Metabolism*, ed. C. A. Neyra, Florida: CRC Press.

Brown, S. A. & Neish, A. C. (1954). The biosynthesis of cell wall carbohydrates: glucose C14 as a cellulose precursor in wheat plants. *Can. J. Biochem. Physiol.* **32**, 170–7.

Chevalier, P. & Lingle, S. E. (1983). Sugar metabolism in developing kernels of wheat and barley. *Crop. Sci.* **23**, 272–7.

Clark, M. G., Bloxham, D. P., Holland, P. C. & Lardy, H. A. (1973). Estimation of the fructose diphosphatase-phosphofructokinase substrate cycle in the flight muscle of *Bombus affinis. Biochem. J.* **134**, 589–97.

Dancer, J. E. & ap Rees, T. (1989). Phosphoribosyl pyrophosphate and the measurement of inorganic pyrophosphate in plant tissues. *Planta* **177**, 261–4.

Doehlert, D. C. (1987). Ketose reductase activity in developing maize endosperm. *Plant Physiol.* **84**, 830–4.

Doehlert, D. C. & Felker, F. C. (1987). Characterisation and distribution of invertase activity in developing maize (*Zea mays*) kernels. *Physiol. Plant* **70**, 51–7.

Doehlert, D. C., Kuo, T. M. & Felker, F. C. (1988). Enzymes of sucrose and hexose metabolism in developing kernels of two inbreds of maize. *Plant Physiol.* **86**, 1013–19.

Echeverria, E., Boyer, C. D., Thomas, P. A., Liu, K.-C. & Shannon, J. C. (1988). Enzyme activities associated with maize kernel amyloplasts. *Plant Physiol.* **86**, 786–92.

Edelman, J., Ginsburg, V. & Hassid, W. Z. (1955). Conversion of monosaccharides to sucrose and cellulose in wheat seedlings. *J. Biol. Chem.* **213**, 843–54.

Entwistle, G. A. & ap Rees, T. (1988). Enzymic capacities of amyloplasts from wheat (*Triticum aestivum*) endosperm. *Biochem. J.* **255**, 391–6.

Gross, P. & ap Rees, T. (1986). Alkaline inorganic pyrophosphatase and starch synthesis in amyloplasts. *Planta* **167**, 140–5.

Hanft, J. M. & Jones, R. J. (1986). Kernel abortion in maize. I. Carbohydrate concentration patterns and acid invertase activity of maize kernels induced to abort *in vitro*. *Plant Physiol.* **81**, 503–10.

Heber, U. (1974). Metabolite exchange between chloroplasts and cytoplasm. *Ann. Rev. Plant Physiol.* **25**, 393–421.

Heber, U. & Heldt, H. W. (1981). The chloroplast envelope:structure, function and role in leaf metabolism. *Ann. Rev. Plant Physiol.* **32**, 139–68.

Higgins, T. J. C. & Easterby, J. S. (1976). Wheatgerm hexokinase: physical and active site properties. *Eur. J. Biochem.* **45**, 145–60.

Huber, S. C. & Akazawa, T. (1986). A novel sucrose synthase pathway for sucrose degradation in cultured sycamore cells. *Plant Physiol.* **81**, 1008–13.

Jenner, C. F. (1976). Wheat grains and spinach leaves as accumulators of starch. In *Transport and Transfer Processes in Plants*, ed. J. B. Passioura, pp. 73–83. New York: Academic Press.

Journet, E. P. & Douce, R. (1985). Enzymic capacities of purified cauliflower bud plastids for lipid synthesis and carbohydrate metabolism. *Plant Physiol.* **79**, 458–67.

Katz, J., Landau, B. R. & Bartsch, G. E. (1966). The pentose cycle, triose phosphate isomerisation, and lipogenesis in rat adipose tissue. *J. Biol. Chem.* **241**, 727–40.

Keeling, P. L., Wood, J. R., Tyson, R. H. & Bridges, I. G. (1988). Starch biosynthesis in developing wheat grain: evidence against the direct involvement of triose phosphates in the metabolic pathway. *Plant Physiol.* **87**, 311–19.

Kumar, R. & Singh, R. (1984). Levels of free sugars, intermediate metabolites and enzymes of sucrose-starch conversion in developing grains. *J. Agric. Food Chem.* **32**, 806–8.

Liedvogel, B. & Kleinig, H. (1980). Phosphate translocator and adenylate translocation in chromoplast membranes. *Planta* **150**, 170–3.

Liu, T.-T. & Shannon, J. C. (1981). Measurement of metabolites associated with non-aqueously isolated starch granules from immature *Zea mays* L. endosperm. *Plant Physiol.* **67**, 525–9.

MacDonald, F. D. & ap Rees, T. (1983). Enzymic properties of amyloplasts from suspension cultures of soybean. *Biochim. Biophys. Acta* **755**, 81–9.

MacLachlan, G. A. & Porter, H. K. (1959). Replacement of oxidation by light as the energy source for glucose metabolism in tobacco leaf. *Proc. Roy. Soc. B.* **150**, 460–73.

Matters, G. L. & Boyer, C. D. (1982). Starch synthases and starch branching enzymes from cotyledons of smooth and wrinkled seeded lines of *Pisum sativum* L. *Biochem. Genet.* **20**, 833–9.

Miernyk, J. A. & Dennis, D. T. (1983). The incorporation of glycolytic intermediates into lipids by plastids isolated from the developing endosperm of castor oil seeds (*Ricinus comunis*. L.). *J. Exp. Bot.* **34**, 712–18.

Muhlethaler, K. (1971). The construction of plastids. In *Structure and Function of Chloroplasts*, ed. M. Gibbs, pp. 7–34. Berlin: Springer-Verlag.

Neish, A. C. (1955). The biosynthesis of cell wall carbohydrates. II. Formation of cellulose and xylan from labelled monosaccharides in wheat plants. *Can. J. Biochem. Phys.* **33**, 658–66.

Robinson, S. P. & Wiskich, J. T. (1977). Pyrophosphate inhibition of carbon dioxide fixation in isolated pea chloroplasts by uptake in exchange for endogenous adenine nucleotides. *Plant Physiol.* **59**, 422–7.

Rognstad, R. & Katz, J. (1980). Control of glycolysis and lipogenesis in the liver by glucagon at the phosphofructokinase–fructose 1,6-bis-phosphate site. *Arch. Biochem. Biophys.* **203**, 62–6.

Seegmiller, C. G., Axelrod, B. & McCready, R. M. (1955). Conversion of glucose-1-C^{14} to pectin in the boysenberry. *J. Biol. Chem.* **217**, 765–75.

Seegmiller, C. G., Jang, R. & Mann, W. Jr. (1956). Conversion of radioactive hexoses to pectin in the strawberry. *Arch. Biochem. Biophys.* **61**, 422–30.

Shafizadey, F. & Wolfram, M. L. (1955). Biosynthesis of C14-labelled cotton cellulose from D-Glucose-1-C14 and D-Glucose-6-C14. *J. Am. Chem. Soc.* **77**, 5182–3.

Shannon, J. C. (1972). Movement of ^{14}C-labelled assimilates into kernels of *Zea mays* L. I. Pattern and rate of sugar movements. *Plant Physiol.* **49**, 198–202.

Shannon, J. C. & Dougherty, C. T. (1972). Movement of ^{14}C-labelled assimilates into kernels of *Zea mays* L. II. Invertase activity of the pedicel and placentochalazal tissues. *Plant Physiol.* **49**, 203–6.

Shannon, J. C. & Garwood, D. L. (1985). Genetics and physiology of starch development. In *Starch, Chemistry and Technology*, pp. 25–86. New York: Academic Press.

Smith, A. M. (1988). Major differences in isoforms of starch-branching enzyme between developing embryos of round- and wrinkled-seeded peas (*Pisum sativum* L.). *Planta* **175**, 270–9.

Turner, J. F. (1969). Starch synthesis and changes in UDPG pyrophos-phorylase and ADPG pyrophosphorylase in the developing wheat grain. *Aust. J. Biol. Sci.* **22**, 1321–7.

Tyson, R. H. & ap Rees, T. (1988). Starch synthesis by isolated amyloplasts from wheat endosperm. *Planta* **175**, 33–8.

Wager, H. G. (1983). The labelling of glucose-6-phosphate in pea slices supplied with ^{14}C glucose and the assessment of the respiratory pathways in the air in 10% CO_2 in N. *J. Exp. Bot.* **34**, 211–20.

R. DOUCE, R. BLIGNY, D. BROWN,
A-J. DORNE, P. GENIX AND C. ROBY

Autophagy triggered by sucrose deprivation in sycamore (*Acer pseudoplatanus*) cells

During the day if nothing inhibits or slows down the rate of photosynthesis almost all plant cells are flooded with sucrose via the phloem. Under these conditions, the immediate products formed in the cytosolic compartment are hexose and UDP-glucose (ap Rees, 1987). These, in turn, are the substrates for a large number of reactions involved in general metabolism and the synthesis of storage products (sucrose and organic acids in the vacuole, starch in the plastids, etc.) as well as for respiration (Douce, 1985). However, during the night or after a long period of drought leading to the closing of stomatal apertures, most of the plant cells are deprived of sucrose.

Work focused on these aspects of sucrose starvation is reported here with heterotrophically grown sycamore cell cultures. We present evidence which shows that after a long period of sucrose starvation, when almost all the intracellular carbohydrate pool has disappeared, cytoplasm can be utilized as a carbon source for respiratory purposes. This autophagic process is distinct from senescence, i.e. the series of events subjected to direct genetic control and concerned with cellular disassembly in the leaf and the mobilization of various small molecules (amino acids, asparagine, etc.) released during this process. This has been excellently reviewed by Thomas & Stoddart (1980), Thomson & Whatley (1980) and Laurière (1983).

Effect of sucrose starvation on the rate of O_2 consumption by sycamore cells

The rate of respiration of cells deprived of sucrose appeared to be constant for at least 24 h (Fig. 1). Thus by a gradual mobilization of starch reserves, the respiratory activity was sustained and stabilized over an extended period of time during which an external carbon source was not available. Thereafter, the rate of O_2 consumption decreased with time. After 50 h of starvation, their capacity to utilize O_2 was reduced to less

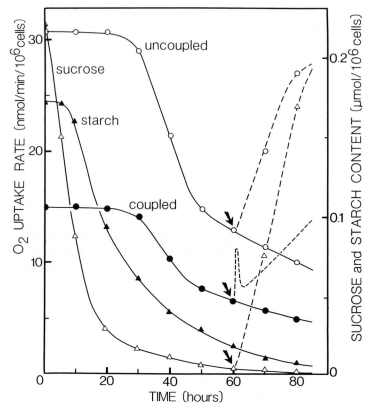

Fig. 1. Effect of sucrose starvation on the rate of O_2 consumption by sycamore cells and on carbohydrate pools (starch and sucrose). Cells harvested from the culture medium were rinsed three times by successive resuspensions in fresh culture medium devoid of sucrose and incubated at zero time (20 mg wet weight per ml) into flasks containing sucrose-free culture medium. At each time cells were harvested and sucrose, starch and the O_2 consumption rate were measured.

The starch content is expressed as μmol glucose per 10^6 cells; ●, normal respiration; ○, uncoupled respiration; the dotted lines correspond to the enhancement of cell respiration and sucrose accumulation that are observed when 50 mM sucrose is added to the culture medium (arrows).

than 50% of that of normal growing cells. Fig. 1 also indicates that the uncoupled rate of O_2 consumption decreased after 30 h in the same ratio as the rate of respiration without uncoupler. A careful examination of Fig. 1 indicates that the rate of O_2 consumption began to decline when the intracellular sucrose had been consumed. Under these conditions starch content was reduced to less than 30% of that of normal cells.

The progressive loss of respiratory capacity of sycamore cells during their ageing in sucrose-free culture medium could be entirely attributable to a progressive decrease in the intracellular substrate levels as suggested by several authors (for review see Yemm, 1965). Furthermore, the respiration rate of many plant tissues falls off when they are deprived of a carbon source for a long period of time, and the reduced respiration rates resulting from such starvation can often be elevated by supplying the appropriate sugars (James, 1953). According to Lambers (1985) in the absence of exogenous sucrose or glucose, intracellular respiratory substrates (pyruvate, malate) are no longer 'wastefully' respired via the cyanide-insensitive alternative pathway leading to a progressive decrease in the rate of O_2 consumption. In other words, according to Lambers (1985) the level of respiratory substrate determines the degree to which the cyanide-resistant pathway contributes to respiration. However, this is most unlikely because in the cells the rates of mitochondrial respiration must be rigorously co-ordinated to meet the ATP demand of the cytoplasm (Brand & Murphy, 1987; Dry, Bryce & Wiskich, 1987; Douce & Neuburger, 1989) and it is difficult to imagine that the intracellular substrate level is the parameter which determines the immediate rates of O_2 consumption even after a long period of sucrose starvation. In support of this idea, mitochondria in sucrose-starved cells are in a state between states 4 and 3 since respiration can be increased (to state 3) with uncouplers (Fig. 1) or decreased (to state 4) by inhibiting ATP production with inhibitors such as oligomycin. It is very likely therefore, that the rate of O_2 consumption by sucrose-starved cells still depends upon the rate of ADP delivery to the mitochondria during the course of metabolism (Roby *et al.*, 1987). Furthermore, if one considers that substrate levels have been depleted after a long period of sucrose starvation we should expect addition of exogenous sucrose to increase respiration immediately afterwards. In fact this is not seen. The enhancement of respiration that is observed when sucrose is added to the culture medium is a slow process. The full rate of uncoupled O_2 consumption by 70-h sucrose-starved cells was recovered several hours after the addition of sucrose (Journet, Bligny & Douce, 1986 and Fig. 1). Likewise, Saglio & Pradet (1980) have clearly shown that O_2 uptake declines progressively after excision of maize root tips and that the addition of exogenous sugars induces a slow rise in the respiratory rate up to its original value while the energy charge remains constant. The most likely explanation for these results is that the number of mitochondria/cell declines progressively during the course of sucrose starvation. In support of this suggestion, studies performed with sycamore cells by Journet *et al.* (1986) indicated that during the course of sucrose starvation the respiration rates decreased progressively in the same ratio as the decrease in intracellular cardiolipin or cytochrome aa_3, two specific mitochondrial markers (Bligny & Douce, 1980).

Effect of sucrose starvation on polar lipid fatty acids of sycamore cells

Figure 2 summarizes the effect of sucrose starvation of sycamore cells on intracellular fatty acid content. After 30 h of sucrose starvation, when almost all of the intracellular carbohydrate pool (starch+sucrose) (see Fig. 1) had disappeared, the cell fatty acid content declined progressively. Analysis of cell polar lipids has indicated that a long period of sucrose starvation also induced a progressive disappearance of all the cell phospholipids and galactolipids (Table 1). During the period of the most rapid disappearance of polar lipid fatty acids, the respiratory quotient (RQ) of the cell was as low as 0·7. Such a low RQ indicates that the fatty acids which were released during the breakdown of the endomembranes may be utilized as respiratory substrates through peroxisomal metabolic pathways (Gerhardt, 1986). This observation is reminiscent of respiration in detached leaves kept in continuous darkness for several days (James, 1953) and in maize root tips deprived of sucrose during several hours where a connection between respiration of lipids and low RQ has been verified (Pradet & Raymond, 1983). Interestingly during the course of polar lipid breakdown the total amount of sterols, including steryl glucoside and acylated steryl glucoside, appeared to be constant (Table 1). These observations confirm that a large proportion of the endomembrane system disappears in starving cells, whilst tonoplast and plasmalemma, enriched in sterol compounds (Hartmann & Benveniste, 1987) escape the autophagic process.

These data raise questions about the mechanism involved in the triggering of this autophagic process. It is possible that ubiquitin, a highly conserved protein involved in several important regulatory processes through its ATP-dependent, covalent ligation to a variety of eukaryotic target proteins or cell membranes (cell surface recognition) (Finley & Varshavsky, 1985; Vierstra, 1987), could play an important role in this energy-dependent autophagic process. It is also possible that tubes of smooth endoplasmic reticulum (ER) wrap themselves around portions of cytoplasm including cell organelles such as mitochondria. Once the sequestered portions of the cytoplasm have been completely enclosed by the lateral fusion of wrapping ER, various hydrolases are probably

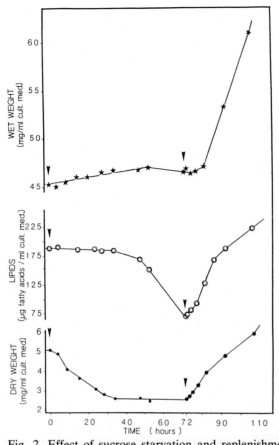

Fig. 2. Effect of sucrose starvation and replenishment on cell weight (wet and dry weight) and on intracellular fatty acid content of sycamore cells. Cells harvested from the culture medium were rinsed three times by successive resuspension in fresh culture medium devoid of sucrose and incubated at zero time (45 mg wet weight ml^{-1}) in flasks containing sucrose-free culture medium. At intervals, cells were harvested on a fibreglass filter (15 s; pressure of suction, 0·2 bar) to ascertain cell wet weight. Cell dry weight was measured after lyophilization of the preceding samples. At arrow 50 mM sucrose was added to the culture medium. The complete oxidation of one glucose molecule requires 6 O_2 (Beevers, 1961) and sycamore cells exhibited an O_2 uptake rate of 42 nmol O_2 h^{-1} mg^{-1} wet weight (Rébeillé *et al.*, 1983). From the knowledge of both these parameters one can calculate that 1·25 µg glucose h^{-1} mg^{-1} wet weight was required in order to sustain the respiration rate of sycamore cells. This value matches the measured rate of dry weight diminution (1·5 µg h^{-1} mg^{-1} wet weight).

Table 1. *Biochemical characterization of normal and sucrose-starved cells*

Cells	Protein mg·10^{-6} cells	Phospholipids	Galactolipids µg·10^{-6} cells	Sterols
Normal	0·53	50	4·5	3·5
Sucrose-starved	0·27	20	2·5	3·7

These data are from a representative experiment and have been reproduced at least three times. Normal cells were harvested from the culture medium. Sucrose-starved cells: normal cells were rinsed three times by successive resuspension in fresh culture medium devoid of sucrose and incubated during 50 h into flasks containing sucrose-free culture medium. Galactolipids: monogalactosyldiacylglycerol+digalactosyldiacylglycerol. Sterols: free sterols+sterylglucoside+acylated sterylglucoside.

released from the surrounding exoplasmic space into the sequestered cytoplasm to form an autophagic vesicle. This phenomenon is reminiscent of the formation of vacuoles in parenchyma cells of root meristems (for review see Marty, Branton & Leigh, 1980).

^{31}P NMR and sucrose deprivation in sycamore cells

Suspension cultures of isolated plant cells are excellent for monitoring metabolism by ^{31}P NMR. For this purpose it is important to be able to obtain easily large quantities of cells under rigorously defined culture conditions. Different methods have been described for cultivating higher plant cells in liquid medium (for review see Bligny & Legnay, 1987). For example, the best time to harvest large amounts of sycamore cells in good physiological state is about 2 days before the end of the exponential phase of growth (Bligny & Legnay, 1987). In order to obtain an intense signal, NMR experiments on isolated plant cells require a densely packed suspension (i.e. to sequester cells within a limited volume approximating that of the NMR receiver coil). However, the problems associated with such dense suspensions of cells include hypoxia, rapid consumption of nutrients, and maintenance of sterility. This has been solved by using an experimental arrangement that enables ^{31}P-NMR spectra of packed plant cells to be recorded while circulating a well-oxygenated medium through the packed suspension under controlled conditions, which simulate as closely as possible the normal physiological state. This technique has been used to study the biochemical effects of sucrose starvation of sycamore cells. The theory of NMR methods and relevant experimental techniques for plant tissues are largely discussed in excellent review articles (Roberts & Jardetzky, 1981; Ratcliffe, 1987; Pfeffer & Gerasimowicz, 1989).

The Apparatus

For long-term experiments (more than three days), all glassware and the medium are sterilized by autoclaving before use. Cells are compressed by hand between two circular perforated Teflon plates from a volume of 150 ml (2×10^6 cells ml^{-1}) to a volume of 20 ml (15×10^6 cells ml^{-1}) fitting exactly into a standard 25-mm NMR tube and carrying (through the centre of the plates) a narrow-bore glass tube (inlet tube), the lower end of which almost reaches the flat bottom of the NMR tube (Fig. 3). A reference capillary, approximately 1·5-mm outer diameter, containing 180 mM methylene diphosphonate (pH 8·9 in 30 mM Tris) is inserted inside the inlet tube along the symmetry axis of the cell sample. A plastic screw cap adapted to the 25-mm NMR tube is used to support the inlet tube and two short glass tubes (output tube and safety output tube). A peristaltic pump is connected to these three tubes (inlet and exit tubes) and the nutrient medium is pumped through the system under slight pressure, that is, the liquid (saturated with O_2) passes slowly through the circular plates and the compressed cells (Fig. 3). The flow rate used routinely (30 ml^{-1} min^{-1}) provides more than 10 times the O_2 consumed by the respiration of the packed sycamore cells at 25 °C. Anoxia is therefore continuously avoided. Such a system prevents gas bubbles from collecting in the cell mass which would otherwise decrease the homogeneity of the magnetic field. We have verified that under these conditions cells can survive at least seven days using a sterile sucrose-containing nutrient medium.

By modifying the composition of the circulating medium, it is possible to perturb the cell metabolism and to monitor the spectral changes simultaneously, for example, to obtain several successive spectra from the same sample.

^{31}P NMR of compressed cells

^{31}P NMR spectra obtained from compressed sycamore cells under aerobic conditions and at pH 6·5 showed two distinct peaks of intracellular P$_i$ (Cyt-P$_i$ and Vac-P$_i$, Fig. 4) at approximately 2·2 and 0·4 ppm, equivalent to approximately pH 7·5 and 5·9, respectively (Roby et al., 1987). These values reflect the presence of the vacuolar P$_i$ pool at the acidic pH and the

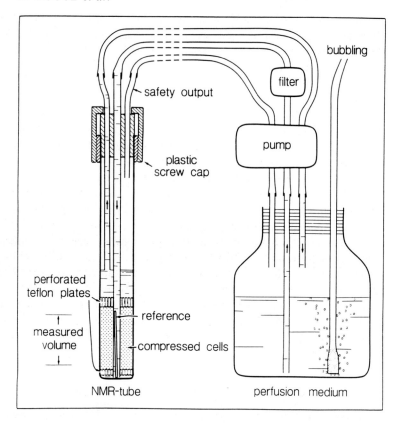

Fig. 3. Apparatus for circulating a nutrient medium through compressed plant cells in a 25 mm (diameter) NMR tube. Cells (about 9 g) were slightly compressed between two circular perforated Teflon plates (measured volume: 16 ml). The cells comprised about 50% of the total volume.

cytoplasmic P_i pool[1] at the slightly alkaline pH (Roberts *et al.*, 1980). The intensities of P_i peaks depend on the relative volumes of cytoplasm and vacuole in the sample.

In the spectra obtained from sycamore cells, no other P_i peaks can be identified unambiguously. This means either that the P_i amounts present

[1]To determine accurately the internal cytoplasmic P_i concentration a calibration of the peak intensity of the P_i resonance with known amounts of external P_i was first performed (data not shown). The curve thus generated gave estimates of cytoplasmic P_i levels of 0·8–1·2 mM including P_i present in the cytosol and various cell organelles. Pfeffer *et al.* (1986) have recently reported a value of 0·7 mM in the case of maize root tips. Under anaerobic conditions this cytoplasmic P_i concentration rises dramatically up to 6 mM (data not shown, see Rébeillé *et al.*, 1983).

in the mitochondrial matrix and the amyloplast stroma of the sycamore cells are below the level of sensitivity of NMR and/or that the pH difference between those organelles and the cytoplasm is too small for the signals to be discriminated. In support of the former suggestion, in sycamore cells the mitochondria comprises only about 10% of the cytoplasmic volume. Furthermore, it has been shown that in plant mitochondria the proton-motive force arises mainly from the membrane potential (250 mV) owing to the presence of a powerful K^+/H^+ antiporter (for a review see Douce, 1985); that is, the ΔpH across the inner mitochondrial membrane during respiration is rather small.

There were also signals from cytoplasmic glucose 6-phosphate, β- and α-UDPG (with contribution of some NAD(P)(H) to the β-peak of UDP-glucose) and γ-, α-, and β-phosphorus of NTP. The region of the spectrum to the high-field side of glucose 6-phosphate (peak *a*) (3·7 ppm) originated from phosphomonoesters. The resonance at 3·3 ppm (peak *b*) could be attributable to NMP and to phosphorylcholine. As the β-NTP peak intensity (the intracellular NTP concentration is approximately 0·25 µmol g^{-1} wet weight) is approximately equal to that of γ-NTP peak (which has some small contribution from β-NDP, Fig. 4) we may assume that *in vivo* the NTP/NDP (free-nucleotides) ratio is high. These data clearly show that sycamore cells have a very high adenylate energy charge (Pradet & Raymond, 1983).

Effect of sucrose deprivation

Figure 5 illustrates the changes that occur in the ^{31}P NMR spectrum when sycamore cells are deprived of sucrose. In this experiment the cells were maintained for more than 70 h in an oxygenated, continuously circulating, solution (P$_i$-free culture medium) at pH 6·5. Little change occurred during the first 10 h. During this time, as previously noted (Fig. 1), the endogenous sucrose content decreased to 30% of the control value and starch (expressed as glucose) remained constant. Under these conditions, sucrose efflux from the vacuole was rapid enough to maintain an optimum phosphate ester concentration in the cytosol. After 10 h of sucrose starvation when a threshold of intracellular sucrose concentration was attained, the rate of sucrose efflux from the vacuole became a limiting factor for cytosolic glycolysis. Consequently, the glucose 6-phosphate resonance decreased gradually, although it was still detectable albeit with a much reduced intensity after 35 h. During this period there was a marked depletion of sugar nucleotides to 20% of the control value. Figure 5 also indicates that P$_i$ molecules liberated from phosphate ester molecules were slowly expelled into the vacuole where they accumulated. Starch

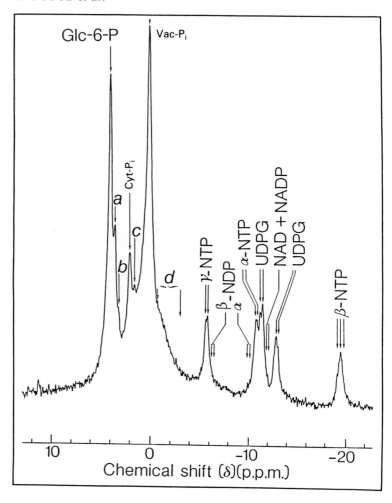

Fig. 4. Representative ³¹P NMR spectrum (81 MHz) of compressed sycamore cells harvested at the end of the exponential phase of growth. Compressed cells (about 8 g) were maintained in a continuously aerated solution (P$_i$- and Mn-free culture medium) and the spectrum was run after 10 h. The spectrum recorded at 15°C is the result of 3000 transients (1 h). Peak assignments: Glc-6-P, glucose 6-phosphate; *a*, position of fructose 6-phosphate, ribose 5-phosphate, and phosphoryl-ethanolamine; *b*, NMP and phosphoryl-choline; cyt-P$_i$, cytoplasmic P$_i$; *c* and *d*, position of glucose 1-phosphate, phosphodiesters and myoinositol hexakisphosphate.

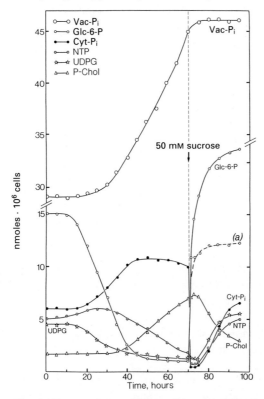

Fig. 5. Time-course evolution of the most abundant ^{31}P NMR-detectable phosphorus compounds in sycamore cells subjected to sucrose starvation for 70 h and followed by 24 h recovery after the addition of 50 mM sucrose and 50 mM P_i (at arrow) to the circulating medium. The concentrations of mobile phosphorus compounds in the cell sample (average over the total sample within the detector) were determined either biochemically or by comparing the area of the signal from methylene diphosphonate capillary reference with the area of the phosphorus resonances obtained from standard solutions of known concentrations of nucleotide triphosphate (NTP); glucose 6-phosphate (Glc-6-P); P_i; UDP-glucose (UDPG), and phosphoryl-choline (P-chol). The broken line (a) shows the recovery of the glucose 6-phosphate peak in the absence of P_i in the circulating medium. Note in this case that the P_i pool does not fluctuate rapidly to buffer the P_i in the cytoplasm. In the 70 h sucrose-starved cells the cytoplasmic P_i and ATP decreased almost immediately upon addition of sucrose, whereas glucose 6-phosphate increased abruptly. Such a situation led to a transient increase of 2-fold in the rate of O_2 consumption ascribed to a marked increase in ADP delivery to the mitochondria (Fig. 1).

mobilization only occurred after 10 h of sucrose starvation (Fig. 1) and it is noteworthy that the lag phase observed for P_i and glucose 6-phosphate ˉchanges in (Fig. 5) was comparable with that observed for starch hydrolysis.

The mechanism of starch breakdown and its conversion to respiratory substrates such as malate and pyruvate is incompletely understood (Steup, 1988). Plastids from soybean cultures (Macdonald & ap Rees, 1983) and cauliflower buds (Journet & Douce, 1985) contain α-glucan phosphorylase and all the enzyme equipment needed to convert glucose 1-phosphate to triose phosphate. These observations strongly suggest that conversion of starch to triose phosphate via glucose 1-phosphate may involve hydrolysis of starch by various amylases and/or phosphorolytic attack, followed by glycolysis to triose phosphate in the stroma and entry into the cytosol via the P_i-translocator (Heldt & Flügge, 1987; Alban, Joyard & Douce, 1988). It is possible that the immediate products of starch breakdown, hexose phosphates (glucose 6-phosphate or glucose 1-phosphate), can move via a specific carrier from the amyloplast to the cytosol. As pointed out judiciously by ap Rees (1987) it is necessary to establish the reactions of starch breakdown and their regulation but also which of the products of starch breakdown (hexose phosphate, free hexose, triose phosphate) move from the amyloplast to the cytosol to sustain mitochondrial respiration during the course of sucrose starvation. The results presented in Figs 1 and 5 strongly suggest that, following sucrose deprivation, an increase in the cytosolic P_i level and a decrease in phosphorylated compounds (glucose 6-phosphate) apparently trigger starch breakdown in amyloplasts. In support of this observation, studies performed with isolated intact chloroplasts also point to a closer correlation between the extraplastidic P_i level and starch mobilization (for review see Steup, 1988).

After 35 h of sucrose deprivation, when the endogenous fatty acids started to decline (Fig. 2), several noticeable changes began to occur in the cells (Fig. 5). Of particular interest was the marked increase in the amount of phosphorylcholine (Fig. 5). This compound was characterized in plants for the first time by Maizel, Benson & Tolbert (1956). It could be demonstrated that the total amount of phosphorylcholine that appeared after a long period of sucrose deprivation corresponded exactly to the total amount of phosphatidylcholine that disappeared within the same period of time (Dorne *et al.*, 1987). Finally curves plotting chemical shift *versus* pH for phosphorylcholine in solutions of various composition indicated that the position of peak b (Fig. 4) corresponds to phosphorylcholine at pH 7·5. This suggests that after a long period of sucrose starvation phosphorylcholine accumulated in the cytoplasmic compartment.

The accumulation of phosphorylcholine which occurs in intact cells during the loss of fatty acids reflects, therefore, the hydrolysis of the major cell membrane polar lipid by a mechanism mediated by lipolytic acyl hydrolase. The fatty acids thus released are probably oxidized in peroxisomes and mitochondria operating in a concerted manner. In support of this suggestion it has been demonstrated that all of the plant cells studied so far contain very powerful lipolytic acyl hydrolase activities (Galliard, 1980) and possess in their peroxisomes all the enzymatic equipment necessary to oxidize free fatty acids (Gerhardt, 1986). Gut & Matile (1988) have found in barley leaf segments that a large proportion of polar lipids was lost during the initial phase of dark-induced senescence and that an apparent induction of key enzymes (isocitrate lyase and malate synthase) of the glyoxylic acid cycle occurred during this period. We can conclude, therefore, that the presence of a large excess of phosphorylcholine in plant cells should be considered as a good marker of membrane utilization after a long period of sucrose starvation and is related to stress.

Effect of sucrose starvation on asparagine accumulation

After 30 h of sucrose starvation, when almost all the intracellular carbohydrate pool (starch+sucrose) had disappeared, the cell protein content declined progressively (see Table 1). A careful analysis of free amino acids carried out during the course of protein breakdown indicated that the total amount of free amino acids which accumulated did not account for all of the total protein which disappeared (Fig. 6). In addition, we observed that ammonia was not released throughout this period. These results strongly suggest that amino acids released during the course of protein breakdown could be metabolized to provide the remaining mitochondria with respiratory substrates. In general, the catabolic sequence for most amino acids leads to a compound which is capable of entering the tricarboxylic acid cycle (Mazelis, 1980). Furthermore the peroxisomes from non-fatty plant tissues are able to activate branched fatty acids which are formed by deamination of leucine, isoleucine and valine, the major amino acids of membrane proteins (Gerbling & Gerhardt, 1988).

These results also suggest that the NH_3 which is released during the course of amino acid utilization is sequestered into organic compounds. For example, it is well established that amides are important products of protein degradation (Lea & Joy, 1983). In support of this last suggestion, Fig. 6 shows that asparagine accumulated considerably in sycamore cells deprived of sucrose. After 70 h of sucrose starvation, the total amount of

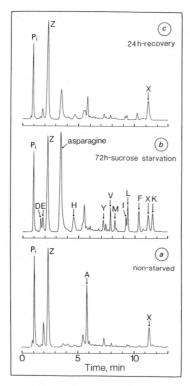

Fig. 6. Time course evolution of the most abundant amino acids in sycamore cells subjected to sucrose starvation for 70 h and followed by 36 h recovery after the addition of 50 mM sucrose to the medium.

(*a*) Separation of amino acids (phenylthiocarbamyl derivatives) from non-starved cells; (*b*) separation of amino acids (phenylthiocarbamyl derivatives) from cells at 72 h following sucrose starvation; (*c*) separation of amino acids (phenylthiocarbamyl derivatives) from cells deprived of sucrose for 72 h and then returned to a sucrose containing medium (50 mM) for 24 h.

The gradient run for the separation consisted of 90% eluent A (0·14 M sodium acetate 4 mM triethylamine, pH 6·4)+10% eluent B [60% (v/v) acetonitrile in water] traversing to 49% eluent A+51% eluent B in 13 min; flow rate 1 ml min^{-1}; column: Pico-Tag analysis column; detector: ultraviolet (254 nm). Note the marked accumulation of asparagine after 70 h sucrose starvation and its rapid disappearance after the addition of sucrose to the medium.

Abbreviations: D, aspartic acid; E, glutamic acid; Z, internal standard (amino 2-adipic acid); H, histidine; U, tyrosine; V, valine; M, methionine; I, isoleucine; L, leucine; F, phenylalanine; X, impurity; K, lysine; P_i, inorganic phosphate; A, alanine.

asparagine present in intact sycamore cells was approximately $1 \, \mu mol \cdot 10^{-6}$ cells and the total amount of protein that had been hydrolysed was approximately $0 \cdot 26 \, mg \cdot 10^{-6}$ cells. From these results we calculated that the total amount of asparagine (expressed as nitrogen) and free amino acids that appeared after a long period of sucrose deprivation corresponded approximately to the total amount of protein (expressed as nitrogen), that disappeared within the same period of time. From these data we conclude that asparagine accumulated in sycamore cells after a long period of sucrose deprivation, that it was derived from the catabolism of amino acids and that asparagine itself was not hydrolysed further and in fact, accumulated in the sucrose-deprived cells.

Many questions remain to be answered regarding the synthesis of asparagine. It might be expected that aspartate aminotransferase, glutamine synthetase, glutamate synthase and asparagine synthetase play a key role in NH_3 reassimilation and in asparagine accumulation (Robertson & Farnden, 1980). Oxaloacetate which is necessary for asparagine biosynthesis, is very likely provided by the tricarboxylic acid cycle. In support of this last suggestion, the catabolic sequence for each amino acid leads to a compound which is capable of entering the tricarboxylic acid cycle (Mazelis, 1980). Furthermore, in various circumstances intact plant mitochondria can export oxaloacetate (Douce & Neuburger, 1989). We can conclude that the presence of a large excess of asparagine in plant cells should be considered as a good marker of protein breakdown after a long period of sucrose starvation and like phosphorylcholine is also likely to be related to stress. In support of this suggestion numerous interesting results (for a review see Stewart & Larher, 1980 and Sieciechowicz, Joy & Ireland, 1988) have clearly indicated that asparagine (along with other amides) is synthesized during exposure of plants to periods of environmental stress such as mineral deficiencies, drought or conditions of increased salinity.

Effect of sucrose replenishment

Addition of sucrose after 70 h of sucrose starvation (i.e. when the uncoupled respiration rate had decreased considerably) (Fig. 1) resulted in a disappearance of the cytoplasmic P_i content and a marked increase in that of glucose 6-phosphate (Fig. 5). Interestingly, although the vacuole had sequestered almost all of the cellular P_i which was liberated during the course of sucrose starvation, the P_i necessary to restart cell metabolism was utilized first from the cytoplasm, and then, if present, from the external medium (Fig. 5). The consumption of vacuolar P_i was a very slow process which required several hours and occurred only in response to the depletion of P_i from the external medium (data not shown): in this series of experiments the vacuolar P_i pool did not fluctuate to buffer the P_i in the cytoplasm. Another interesting feature is that the amount of ATP and cytoplasmic P_i slowly returned to near the level of normal cells. This slow increase in the intracellular amount of metabolites is probably attributable to the fact that synthesis of new cytoplasmic material is a very slow process requiring several hours.

Addition of 50 mM sucrose to the medium after 70 h of sucrose starvation resulted in a marked increase in the cell dry weight (Fig. 2) and total cell fatty acids associated with polar lipids (Fig. 2). The increase in the cell dry weight was attributable to a rapid accumulation of sucrose in the vacuole and starch in plastids (Fig. 1) whereas the increase in total cell fatty acids was attributable to the synthesis of new cytoplasmic material such as mitochondria (Dorne *et al.*, 1987). During this time the cell fresh weight per ml of culture medium remained constant. When the total fatty acid content had returned to near the level of normal cells (Fig. 2), growth began. The lag phase preceding growth represented the time required to synthesize new cytoplasmic material. Furthermore, 10 h after replenishment of sucrose other substantial changes were evident in the cells. Of particular interest was the marked decrease in the amount of phosphorylcholine (Fig. 5). We have shown that the linear decrease in cytosolic phosphorylcholine closely correlated with the linear increase in phosphatidylcholine (Roby *et al.*, 1987). Under these conditions phosphorylcholine was readily metabolized to sustain phosphatidylcholine synthesis, the major polar lipid of all the cell membrane systems (Harwood, 1980).

Our results also indicate that in the presence of sucrose, asparagine was found to be metabolized (Fig. 6) either by transamination or deamidation by an asparaginase, in agreement with what was previously observed in developing pea leaves and seeds (for review see Sieciechowicz *et al.*, 1988). However, in mature leaves that no longer require nitrogen for growth, asparagine is not readily metabolized and is re-exported towards the xylem where it accumulates (for review see Sieciechowicz *et al.*, 1988).

In summary these results demonstrate that during the course of sucrose starvation the catabolism of fatty acids and amino acids must be sufficiently intense to supply mitochondria with respiratory substrates in order to maintain a high nucleotide energy charge in the remaining cytoplasmic fraction. These observations are important if one considers that ATP is a very important factor in controlling the rate of autophagy in rat hepatocytes (Plomp *et al.*, 1987). These results emphasize the extraordinary flexibility and complexity of plant cell metabolism. This flexibility is compounded by the fact that cytoplasm, in particular, can be utilized as a carbon source after a long period of sucrose starvation without significantly affecting the survival of these cells. Under these conditions, plant cells, owing to the presence of intracellular pools of carbohydrate and to their ability to control an autophagic process, can survive for a long period of time without receiving any external supply of organic carbon.

References

Alban, C., Joyard, J. & Douce, R. (1988). Preparation and characterization of envelope membranes from non-green plastids. *Plant Physiol.* **88**, 709–17.

ap Rees, T. (1987). Compartmentation of plant metabolism. In *The Biochemistry of Plants*, vol. 12, ed. D. D. Davies, pp. 87–115. New York: Academic Press.

Beevers, H. (1961). *Respiratory Metabolism in Plants*. Evanston: Row Peterson and Company.

Bligny, R. & Douce, R. (1980). A precise localization of cardiolipin in plant cells. *Biochim. Biophys. Acta* **617**, 254–63.

Bligny, R. & Legvay, J. J. (1987). Techniques of cell suspension culture. *Meth. Enzymol.* **148**, 3–16.

Brand, M. D. & Murphy, M. P. (1987). Control of electron flux through the respiratory chain in mitochondria and cells. *Biol. Rev.* **62**, 141–93.

Dorne, A. J., Bligny, R., Rebeillé, F., Roby, C. & Douce, R. (1987). Fatty acid disappearance and phosphorylcholine accumulation in higher plant cells after a long period of sucrose deprivation. *Plant Physiol. Biochem.* **25**, 589–95.

Douce, R. (1985). *Mitochondria in Higher Plants: Structure, Function and Biogenesis.* New York: Academic Press.

Douce, R. & Neuburger, M. (1989). The uniqueness of plant mitochondria. *Ann. Rev. Plant Physiol.* **40**, 371–414.

Dry, J. B., Bryce, J. H. & Wiskich, J. T. (1987). Regulation of mitochondrial respiration. In *The Biochemistry of Plants*, vol. 11, ed. D. D. Davies, pp. 213–52. New York: Academic Press.

Finley, D. & Varshavsky, A. (1985). The ubiquitin system: functions and mechanisms. *Trends Biochem. Sci.* ed. Pers., **10**, 343–7.

Galliard, T. (1980). Degradation of Acyl Lipids: Hydrolytic and Oxidative Enzymes. In *The Biochemistry of Plants*, vol. 4, ed. P. K. Stumpf & E. E. Conn, pp. 85–116. New York: Academic Press.

Gerbling, H. & Gerhardt, B. (1988). Oxidative decarboxylation of branched-chain 2-oxo-fatty acids by higher plant peroxisomes. *Plant Physiol.* **88**, 13–15.

Gerhardt, B. (1986). Basic metabolic function of the higher plant peroxisome. *Physiol. Vég.* **24**, 397–410.

Gut, H. & Matile, P. (1988). Apparent induction of key enzymes of the glyoxylic acid cycle in senescent barley leaves. *Planta* **176**, 548–50.

Hartmann, M. A. & Benveniste, P. (1987). Plant Membrane sterols: isolation, identification and biosynthesis. *Meth. Enzymol.* **148**, 632–50.

Harwood, J. L. (1980). Plant acyl lipids: structure, distribution and analysis. In *The Biochemistry of Plants*, vol. 4, ed. P. K. Stumpf & E. E. Conn, pp. 1–55. New York: Academic Press.

Heldt, H. W. & Flügge, V. I. (1987). Subcellular transport of metabolites in plant cells. In *The Biochemistry of Plants*, vol. 12, ed. D. D. Davies, pp. 49–85. New York: Academic Press.

James, W. O. (1953). *Plant Respiration*. Oxford: Clarendon Press.

Journet, E. & Douce, R. (1985). Enzymatic capacities of purified cauliflower bud plastids for lipid synthesis and carbohydrate metabolism. *Plant Physiol.* **79**, 458–67.

Journet, E., Bligny, R. & Douce, R. (1986). Biochemical changes during sucrose deprivation in higher plant cells. *J. Biol. Chem.* **261**, 3193–9.

Lambers, H. (1985). Respiration in intact plants and tissue: its regulation and dependence on environmental factors, metabolism and invaded organisms. In *Encyclopedia of Plant Physiology*, vol. 18, ed. R. Douce & D. D. Day, pp. 418–73. Berlin: Springer-Verlag.

Laurière, C. (1983). Enzymes and leaf senescence. *Physiol. Vég.* **21**, 1159–97.

Lea, P. J. & Joy, K. W. (1983). Amino acid interconversion in germinating seeds. In *Recent Advances in Phytochemistry*, vol. 17, ed. C. Nozzdilo, P. J. Lea & E. Loewus, pp. 77–95. New York: Plenum Press.

Macdonald, F. D. & ap Rees, T. (1983). Enzymic properties of amyloplasts from suspension cultures of soybean. *Biochim. Biophys. Acta* **755**, 81–9.

Maizel, J. V., Benson, A. A. & Tolbert, N. E. (1956). Identification of phosphorylcholine as an important constituent of plant saps. *Plant Physiol.* **31**, 407–8.

Marty, F., Branton, D. & Leigh, R. A. (1980). Plant Vacuoles. In *The Biochemistry of Plants*, vol. 1, ed. N. E. Tolbert, pp. 625–58. New York: Academic Press.

Mazelis, M. (1980). Amino Acid Catabolism. In *The Biochemistry of Plants*, vol. 5, ed. B. J. Miflin, pp. 541–67. New York: Academic Press.

Pfeffer, P. E., Tu, S. I., Gerasimowicz, M. V. & Cavanagh, J. R. (1986). *In vivo* ^{31}P NMR studies of corn root tissue and its uptake of toxic metals. *Plant Physiol.* **80**, 77–84.

Pfeffer, P. E. & Gerasimowicz, W. V. (1989). In *Nuclear Magnetic Resonance in Agriculture*, ed. P. E. Pfeffer & W. V. Gerasimowicz, pp. 3–70. Boca Raton: CRC Press.

Plomp, P. J. A. M., Wolvetang, E. J., Groen, A. K., Meijer, A. J., Gordon, P. B. & Seglen, P. O. (1987). Energy dependence of autophagic protein degradation in isolated rat hepatocytes. *Eur. J. Biochem.* **164**, 197–203.

Pradet, A. & Raymond, P. (1983). Adenine nucleotide ratios and adenylate energy charge in energy metabolism. *Ann. Rev. Plant Physiol.* **34**, 199–224.

Ratcliffe, R. G. (1987). Application of nuclear magnetic resonance methods to plant tissues. *Meth. Enzymol.* **148**, 683–700.

Rebeillé, F., Bligny, R., Martin, J. B. & Douce, R. (1983). Relationship between the cytoplasm and the vacuole phosphate pool in *Acer pseudoplatanus* cells. *Arch. Biochem. Biophys.* **225**, 143–8.

Roberts, J. K. M. & Jardetzky, O. (1981) Monitoring of cellular metabolism by NMR. *Biochim. Biophys. Acta* **639**, 53–76.

Roberts, J. K. M., Ray, P. M., Wade-Jardetzky, N. & Jardetzky, O. (1980). Estimation of cytoplasmic and vacuolar pH in higher plant cells by ^{31}P NMR. *Nature* **283**, 870–2.

Robertson, J. G. & Farnden, K. J. F. (1980). Ultrastructure and metabolism of the developing legume. In *The Biochemistry of Plants*, vol. 5, ed. B. J. Miflin, pp. 65–113. New York: Academic Press.

Roby, C., Martin, J. B., Bligny, R. & Douce, R. (1987). Biochemical changes during sucrose deprivation in higher plant cells. ^{31}Phosphorus nuclear magnetic resonance studies. *J. Biol. Chem.* **262**, 5000–7.

Saglio, P. H. & Pradet, A. (1980). Soluble sugars, respiration, and energy charge during aging of excised maize root tips. *Plant Physiol.* **66**, 516–19.

Sieciechowicz, K. A., Joy, K. W. & Ireland, R. J. (1988). The metabolism of asparagine in plants. *Phytochemistry*, 27, 663–71.

Steup, M. (1988). Starch degradation. In *The Biochemistry of Plants*, vol. 14, ed. J. Preiss, pp. 255–96. New York: Academic Press.

Stewart, G. R. & Larher, F. (1980). Accumulation of amino acids and related compounds in relation to environmental stress. In *The Biochemistry of Plants*, vol. 5, ed. B. J. Miflin, pp. 609–35. New York: Academic Press.

Thomas, H. & Stoddart, J. L. (1980). Leaf senescence. *Ann. Rev. Plant Physiol.* **31**, 83–111.

Thomson, W. W. & Whatley, J. M. (1980). Development of non-green plastids. *Ann. Rev. Plant Physiol.* **31**, 375–94.

Vierstra, R. D. (1987). Demonstration of ATP-dependent, ubiquitin conjugating activities in higher plants. *Plant Physiol.* **84**, 332–6.

Yemm, E. W. (1965). The respiration of plants and their organs. In *Plant Physiology. A treatise*, vol. 4A, ed. F. C. Steward, pp. 231–310. New York: Academic Press.

M. J. EMES AND C. G. BOWSHER

Integration and compartmentation of carbon and nitrogen metabolism in roots

Introduction

The understanding of nitrate assimilation in higher plants is a subject which is constantly being reviewed and updated because of its central importance in understanding the relationship between plant metabolism and growth and because of its agricultural significance. There have been a number of such reviews of this subject in recent years (e.g. Beevers & Hageman, 1983) though inevitably emphasis has been on the leaf system since that is better understood. Recently Oaks & Hirel (1985) produced an excellent account of nitrogen metabolism in roots. The aim of this chapter is to build on that foundation and to lay particular emphasis on the intracellular compartmentation of the events and pathways which make up nitrate assimilation in roots.

In lowland agricultural soils, nitrate is the predominant form of inorganic nitrogen available to plants and can reach 20 mM in areas where fertilizer has been applied (Russell, 1973). In natural soils where nitrate is present the concentration may be less than 1 mM and in some acidic soils it may be completely absent. Just as there is variation in the availability of nitrate for use by higher plants so there is divergence between those plants which assimilate nitrate mainly in the leaf and those which are root assimilators. Andrews (1986) and Smirnoff & Stewart (1985) have summarized the information on root and shoot nitrate assimilators. The main conclusions appear to be that in a wide range of temperate annual and perennial species, root nitrate assimilation is more important when growing in low external nitrate concentrations and that shoot assimilation becomes more important as external nitrate concentration increases. It appears that for tropical and subtropical species assimilation of nitrate is predominantly in the shoot at low and high external NO_3^--concentrations. It must be emphasized, however, that within any of the above classifications (i) there is considerable variation between species in the proportions of root: shoot assimilation, (ii) the proportion of assimilation taking place

147

in each tissue is a function both of external concentration of nitrate and of tissue age, and (iii) estimates are often based on '*in vivo*' nitrate reductase assays rather than actual measurement of the amounts of nitrate assimilated. Further, the time of sampling is important. For example in 8-day old barley seedlings the proportion of assimilation taking place in the root doubles from 20 to 40% when comparing plants in light and dark (Aslam & Huffaker, 1982).

Pathway of nitrate assimilation

The enzymes involved in the assimilation of inorganic nitrate and the primary formation of glutamate are nitrate reductase (EC 1.6.6.1), nitrite reductase (EC 1.7.7.1), glutamine synthetase (EC 6.3.1.2) and either ferredoxin-dependent glutamate synthase (EC 1.4.7.1) or NADH-dependent glutamate synthase (EC 1.4.1.14) (Fig. 1). The first two of

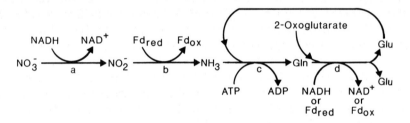

Fig. 1. Pathway of nitrate assimilation in higher plants. a, nitrate reductase; b, nitrite reductase; c, glutamine synthetase; d, glutamate synthase. (Fd=ferrodoxin in reduced or oxidized form).

these enzymes are induced by the addition of NO_3^- (Beevers & Hageman, 1983). Nitrate reductase (NR) is NADH dependent in roots of most species although there have been reports of NADPH dependent activity in corn roots (Redinbaugh & Campbell, 1981). The enzyme has been studied far more extensively in leaf tissue. With the cloning of the NR gene (Calza *et al.*, 1987; Cheng *et al.*, 1986; Crawford *et al.*, 1986) it has been shown that upon addition of nitrate, induction occurs at the transcriptional level with the NR mRNA levels increasing considerably. Recently Galangeau *et al.* (1988) have looked at the synthesis of the NR mRNA in tomato and tobacco leaves. They found that the peak in NR mRNA precedes NR synthesis and that mRNA increases in the dark period but decreases in the light. Conversely, NR protein levels increased in the light but the activity of the enzyme did not follow the NR protein profile closely. Rothstein and co-workers (Kramer *et al.*, 1989; Privalle *et al.*, 1989) have recently examined maize root, leaf and cell suspensions and found that the message for nitrite reductase is induced transiently and is short-lived in roots with a half-life of less than 30 minutes. The level of nitrite reductase protein in roots however, continues to increase long after the peak of mRNA has passed and continues to rise up to 24 h after the application of nitrate. There is clearly substantial post transcriptional control of nitrite reductase (and by analogy with leaves, NR) synthesis in roots which merits considerable attention if we are to understand the control of enzyme synthesis properly.

There is now little doubt that the primary assimilation of ammonia in roots is via the glutamine synthetase/glutamate synthase pathway (Miflin & Lea, 1980; Fentem, Lea & Stewart, 1983). Because of its central position with respect to primary ammonia assimilation, photorespiratory reassimilation and nitrogen fixation, isoenzymes of glutamine synthetase (GS) have been extensively studied in leaves, roots and nodules of higher plants (McNally & Hirel, 1983; Lightfoot, Green & Cullimore, 1988; Tingey, Walker & Coruzzi, 1987). Two novel forms of GS are produced in the plant fraction of legume root nodules during nodulation (Cullimore *et al.*, 1983). In non-nodulated or non-leguminous plants, however, the weight of evidence suggests that there is only one form of GS present in roots (Mann, Fentem & Stewart, 1979; McNally & Hirel, 1983). There is to date one report of two isoenzymes of GS being separated from roots of pea and alfalfa by DEAE chromatography (Vezina *et al.*, 1987), and the significance of this will be discussed in relation to the intracellular locations of these enzymes shortly.

Glutamate synthase has been less extensively studied than glutamine synthetase and there are few reports in the literature of its purification from roots (Boland & Benny, 1977; Chen & Cullimore, 1988). There are at least two forms of glutamate synthase in most roots, one which is NADH dependent and the other which utilizes reduced ferredoxin as a cofactor and is immunologically distinguishable from its leaf counterpart (Suzuki, Vidal & Gadal, 1982). Recently two additional forms of glutamate synthase have been identified in the plant fraction of *Phaseolus vulgaris* root nodules (Chen & Cullimore, 1988), and again these appear to be synthesized in response to nodulation and are involved in amino acid synthesis in this organ.

Compartmentation of nitrogen assimilation in roots

Despite the obvious heterogeneity of higher plant roots there have been few studies of the intercellular distribution of the enzymes of nitrate assimilation. Sarkissian & Fowler (1974) demonstrated that most of the extractable NR from pea roots was in the apical segments of the root. However, they took no special precautions to protect NR from proteolytic digestion or inactivation and these can cause a marked reduction in NR activity *in vitro* (Solomonson *et al.*, 1984). It would seem unlikely that NR is confined solely to the root apices in view of the fact that GS activity increases with distance from the apex (Vezina & Langlois, 1989).

Intracellularly, NR is located in the cytoplasm (Emes & Fowler, 1979a) whilst both nitrite reductase and glutamate synthase are located in root plastids (Miflin, 1974; Emes & Fowler, 1979a) though in lupin NADH-glutamate synthase appears to be cytosolic (Boland & Benny, 1977). The distribution of GS merits some attention as there is conflicting evidence with regard to the number and subcellular distribution of isoenzymes in roots. Earlier studies had shown that this enzyme was associated with the plastid fraction after centrifugation of root homogenates on sucrose density gradients (Miflin, 1974; Emes & Fowler, 1979a). However, by far the greater portion of activity (some 90% or more) was always located in the same fraction as NR (cytosol) and this was shown to be the case for a number of species (Suzuki, Gadal & Oaks, 1981). Separation of chloroplastic and cytosolic leaf glutamine synthetases is readily achieved by DEAE anion exchange chromatography, but only one form was identifiable from root extracts treated in the same way (Mann, Fentem & Stewart, 1979) and this was clearly a cytoplasmic enzyme. A likely interpretation of the earlier results was, therefore, that some GS had become adsorbed to the root plastid preparations during isolation, since there had been no test to see whether this plastidial activity was latent within the organelle. However, a number of recent observations suggest that there may be a genuine GS isoform within these non-photosynthetic plastids. First, Emes & Fowler (1983) demonstrated that GS activity increased in pea root plastids during the induction of the enzymes of nitrate assimilation. This has recently been confirmed by Vezina & Langlois (1989) who have demonstrated the presence of 38 kDa and 44 kDa subunits of GS in pea roots, the former being present only in the cytoplasmic fraction whereas both were found associated with plastids. Both polypeptides were detected with an anti-glutamine synthetase IgG raised against the enzyme from *Phaseolus* root nodules which also recognized the leaf subunits (Cullimore & Miflin, 1984) and their abundance in root plastids increased markedly after pre-treatment of roots with nitrate. Second, the GS activity was found to be latent in root plastids prepared by more careful extraction procedures (Emes & England, 1986). Thirdly, Vezina *et al.* (1987) have demonstrated that there are two isoenzymes of GS in pea and alfalfa roots, which are separable by DEAE chromatography, and that the ratio of synthetase:transferase activity is different for the soluble and particulate enzymes. There seems therefore to be good evidence that in pea roots, at least, there is an organelle specific plastid form of GS, the synthesis of which is co-ordinated with the expression of nitrite reductase within the same organelle and NR in the cytoplasm. If there are genuine species differences and it turns out that in some cases all the GS is cytosolic then this has implications for the movement of metabolites in and out of the root plastid. In leaves, once nitrite has entered the chloroplast it can be reduced and incorporated into amino acids within the same organelle. If there is no plastid GS in roots then it follows that ammonia, produced by nitrite reduction, must then leave the organelle, be assimilated in the cytoplasm and re-enter the organelle as glutamine in order for the glutamate synthase reaction to occur. This is by no means impossible but begs the question as to the significance of the difference between roots and leaves, and necessitates more research on the role of glutamine synthetase in roots. It is perhaps worth noting that photorespiratory mutants of barley leaves which lack chloro-plastic, i.e. plastidial GS grow normally under non-photorespiratory conditions suggesting that it may not be essential for primary nitrogen assimilation in barley (Blackwell, Murray & Lea, 1987; Wallsgrove *et al.*, 1987).

Integration of nitrogen and carbon metabolism in roots

Origin of reductant for nitrate reductase

In order for the reduction of one molecule of nitrate to ammonia and its incorporation into amino acids to take place, a total of 10 electrons and one molecule of ATP are required (Fig. 1). Clearly, in photosynthetic tissues the energy requirements of this process are met directly/indirectly through the photochemical generation of reductant and ATP whereas in roots they would have to be generated via catabolism of carbohydrates. Nance (1948) first commented on the ability of nitrate assimilation to occur in the absence of oxygen. From this it was concluded that the oxidation of carbohydrates, by providing reductants, can be coupled with the nitrate reducing system. An increase in CO_2 production when nitrate was added to barley roots immersed in culture solution was observed by Willis & Yemm (1955). Butt & Beevers (1961) made similar observations with maize roots during the assimilation and reduction of nitrite. Experiments with a split root system showed that fed roots are able to import more carbohydrates from the shoots than roots grown without nitrate (Lambers *et al.*, 1982).

Questions, therefore, arise as to the precise source of reductant, ATP and carbohydrate skeletons necessary for amino acid synthesis in the intact non-green cell; the intracellular location of the 'supply' pathways with respect to the 'demand' enzymes, and the mechanism by which 'supply' is matched with 'demand'.

The pathways of carbohydrate oxidation are glycolysis, the oxidative pentose phosphate pathway (OPPP) and the Krebs tricarboxylic acid (TCA) cycle. Either glycolysis or the Krebs cycle could generate NADH for nitrate reductase. The Krebs cycle enzymes are located in the mitochondrial matrix whilst glycolysis is located both in the cytosol and probably in plastids of roots (see later). NADH for nitrate reductase in theory could be generated in one of three places, though to date only the cytosol and mitochondria have been considered, and I will confine the discussion to these two since there is thus far no evidence to suggest that a plastid located glycolytic pathway has a role in nitrate reduction. Reports by Klepper, Flesher & Hageman (1971) and Mann, Hucklesby & Hewitt (1978) indicated that the triose phosphate dehydrogenase step of glycolysis was the likely source of NADH for NR in leaves, using photosynthate generated in the chloroplast. Such a conclusion was based upon experiments in which leaf tissue was supplied with a range of metabolites (mainly glycolytic intermediates and organic acids) and measurement of '*in vivo*' nitrate reduction. The latter involved use of a freeze-thaw assay, likely to disrupt intracellular membranes to allow penetration of intermediates to the site of the enzymes. The production of nitrite under such conditions was maximal when tissue was supplied with NAD^+ and either glyceraldehyde 3-phosphate or dihydroxyacetone phosphate although malate and isocitrate could support lower rates of nitrite reduction. The fact that NAD^+ was necessary to support malate and isocitrate dependent nitrate reduction (Mann *et al.*, 1978) suggests that considerable disruption of membrane integrity had occurred leaving open the possibility that the results may be artefactual, since the natural organization of the cell and the selective permeability of its membranes had been lost.

As an alternative proposal, Nicholas and co-workers have suggested that NADH for nitrate reduction is derived from malate metabolism (Naik & Nicholas, 1986). This proposal is based upon experiments again supplying Krebs cycle intermediates and using an *in vivo* assay for NR, but without freeze-thawing (Sawhney, Naik & Nicholas, 1978). Other experiments have involved the feeding of inhibitors/uncouplers of the mitochondrial and photosynthetic electron transport chains with a view to manipulating NADH availability from the mitochondria/chloroplast either directly, or through effects on energy charge, though such experiments have to be treated with caution (Gray & Cresswell, 1984). A close relationship between respiratory CO_2 release and nitrate reduction further suggested a link between Krebs cycle turnover and the supply of NADH for nitrate reduction (Naik & Nicholas, 1981). However, much of the debate as to whether malate or triosephosphate metabolism generates NADH for NR is actually concerned with the regulation of leaf NR by light. One argument runs that photosynthetically generated triosephosphate is exported from the chloroplast in the light to support NR (amongst other things) and that this supply is diminished in the dark. Alternatively, proponents of the view that malate oxidation supports NR argue that mitochondrial NADH is diverted from the mitochondria in the light for use by NR following the inhibition of respiratory chain activity in the light. However, there is no conclusive evidence to support the concept of light-mediated inhibition of the respiratory chain (Dry, Bryce & Wiskich, 1987).

There have been few reports concerning the source of electrons for NR in roots. Nitrate reduction is obviously dependent on the oxidation of imported photosynthate, but Aslam & Huffaker (1982) have shown that, in barley roots at least, nitrate reduction in roots proceeds at the same rate in light and dark periods of a growth regime. Naik & Nicholas (1984) have examined the source of NADH for NR in wheat roots. During nitrate assimilation under aerobic conditions, addition of the respiratory uncoupler dinitrophenol (DNP) caused an accumulation of nitrite in the tissue. This was attributed to a lack of ATP for the synthesis of sufficient glucose 6-phosphate to support nitrite reduction via the oxidative pentose phosphate pathway (Dry, Wallace & Nicholas, 1981, and see later), hence nitrite is produced but not removed. In the presence of DNP, the inhibitors of malate metabolism, D-malate and malonate inhibited nitrite accumulation implying that the metabolism of malate, possibly by the mitochondrial malic enzyme, is linked to nitrate reduction. This inhibition of nitrite accumulation by malonate was relieved by supplying fumarate. However, the data is difficult to interpret because of the complex effects of DNP and metabolic inhibitors like malonate. Dancer & ap Rees (1989) have shown that 25 μM DNP (the same concentration as used by Naik & Nicholas) increases the rate of glycolysis in pea roots and *Arum* clubs, and this agrees with the postulate that glucose 6-phosphate is being removed more rapidly in the presence of DNP. The corollary of this is that turnover of triosephosphate in glycolysis will also be increased and, if supplying NADH to NR, would result in an increase in nitrite accumulation. Malonate addition would undoubtedly cause a decrease in TCA cycle activity but whether the impact of this on nitrate reduction is direct or indirect cannot be determined. Malonate would cause an accumulation of organic acids which would feed back to inhibit glycolytic flux (D. T. Dennis, personal comm.) and so diminish NADH production in the cytoplasm. That fumarate relieves malonate inhibition of nitrite accumulation could be because a reduced glycolytic rate may poise the system in such a way that significant amounts of NADH can only be generated under the experimental conditions by supplying an organic acid whose oxidation in the cytosol or mitochondria could support NR. This is not an argument against a role for malate metabolism in supporting NR through reductant supply, but another way of looking at the data. In other words, the case is not proven. However, it is worth noting that pre-incubation of pea roots in nitrate had no effect on the release of $^{14}CO_2$ from [4-^{14}C] malate supplied to this tissue (Fowler & Sarkissian, 1975). This reaction is diagnostic of the operation of malic enzyme and implies that there is no increased flux through malate to meet the demands of nitrate assimilation.

Origin of reductant for nitrite reductase

There has been a longstanding hypothesis that during the induction of nitrate assimilation there is an increased flow of carbon through the OPPP which is linked to the demand for reductant by nitrite reductase (Butt & Beevers, 1961; Sarkissian & Fowler, 1974; Jessup & Fowler, 1977). This is based largely on the results of release of carbon dioxide from carbon atoms 1 and 6 of labelled glucose fed to cells assimilating nitrate. An enhanced release of $^{14}CO_2$ from [1-^{14}C]-glucose above that from [6-^{14}C]-glucose is taken as indicative of OPPP activity and has been shown to increase during nitrate assimilation. However, earlier work took no account of the intracellular compartmentation of events and it has now been demonstrated that the OPPP has a substantial part of its activity associated with non-photosynthetic plastids of roots and other non-green tissues (Emes & Fowler, 1979b; Simcox et al., 1977; Washitani & Sato, 1977) and chloroplasts (Schnarrenburger et al., 1973) as well as being cytoplasmic thereby complicating the interpretation of such data. Accompanying the induction of nitrate assimilation, increases in activity of the plastidic OPPP enzymes glucose 6-phosphate dehydrogenase and 6-phosphogluconate dehydrogenase have been observed (Emes & Fowler, 1983; Emes, Asihara & Fowler, 1979). This implies a spatial and dynamic cooperation between nitrite reductase and the OPPP located within the same cell compartment since it removes the need for transport across the organelle membrane of reducing equivalents and, providing there is sufficient hexose phosphate available for oxidation, ensures a closely coupled supply of electrons. This relationship between hexosephosphate metabolism and nitrite reduction has recently been examined in more detail by Bowsher, Hucklesby & Emes (1989) using purified, intact preparations of pea root plastids. Intact organelles were supplied with nitrite and a number of different metabolites (but no pyridine nucleotides) and the rate of nitrite reduction followed. The most effective substrates supporting nitrite reduction were glucose 6-phosphate, fructose 6-phospate and ribose 5-phosphate, all intermediates of the OPPP. Triose phosphates were ineffective. More detailed kinetic analysis shows that the K_m (G6P) for glucose 6-phosphate dependent nitrite reduction is identical to the K_m of glucose 6-phosphate dehydrogenase for the same substrate (Fig. 2). This could of course merely be a

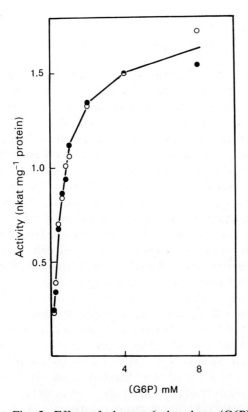

(G6P) mM

Fig. 2. Effect of glucose 6-phosphate (G6P) concentration on G6P-dependent nitrite reduction in pea root plastids (closed circles) and plastidic glucose 6-phosphate dehydrogenase (open circle). Units of activity are nkat mg^{-1} plastid protein. Values for G6P-dependent nitrite reduction are ×17 to allow visual comparison of kinetics.

fortuitous coincidence, since the concentrations of glucose 6-phosphate used to support nitrite reduction in these preparations is varied *outside* the organelle and we do not know what implications this has for the concentration of glucose 6-phosphate inside the plastids. Nevertheless, such results offer strong circumstantial evidence that the oxidation of glucose 6-phosphate is linked to nitrite reduction. Recently Borchert, Grosse & Heldt (1989) have looked at the kinetics of glucose 6-phosphate uptake by pea root plastids and their data imply that glucose 6-phosphate is taken up by the phosphate translocator in these organelles (Emes & Traska, 1987) with a K_m (glucose 6-phosphate) of between 0·1 and 0·5 mM. Their results show that inorganic phosphate and glucose 6-phosphate can counter exchange across the plastid envelope. Surprisingly glucose 1-phosphate did not counter exchange with inorganic phosphate or inhibit its uptake suggesting that either it does not enter root plastids (unlikely in view of the observation that glucose 1-phosphate can support nitrite reduction (Bowsher *et al.*, 1989)), or that it enters by a different mechanism/translocator. Oji *et al.* (1985) have also found that glucose 6-phosphate can support nitrite reduction in barley root plastids, but activity was enhanced by $NADP^+$ suggesting that the membranes of these preparations may have been damaged.

Some of the most convincing evidence that the OPPP supports nitrite reduction comes from experiments in which purified root plastids were fed [^{14}C]-glucose 6-phosphate in which different carbon atoms were specifically labelled (Bowsher *et al.*, 1989) and $^{14}CO_2$ evolution followed under different conditions. $^{14}CO_2$ release was stimulated from carbon atom 1 when nitrite was simultaneously added to the organelles, but not from any of the other carbon atoms examined (C-2, 3, 4 and 6). Further, this enhancement by nitrite of CO_2 release from C-1 of glucose 6-phosphate was only apparent with organelles in which nitrite reductase had been induced by the prior treatment of roots with nitrate, and was not observed in plastids lacking nitrite reductase. The release of CO_2 from C-1 of glucose 6-phosphate can be taken as indicative of the subsequent decarboxylation of 6-phosphogluconate by the pentose phosphate pathway as these organelles lack a tricarboxylic acid cycle. If organelles in which nitrite reductase is present are supplied with [1-^{14}C]-glucose 6-phosphate then the rate of release of $^{14}CO_2$ increases with the concentration of nitrite supplied (Fig. 3, and Bowsher *et al.*, 1989). All these results point to the function of the OPPP in supplying reducing power to nitrite reductase in root plastids.

This raises another issue. The oxidation of glucose 6-phosphate by this pathway generates NADPH, but root and leaf nitrite reductases will not use pyridine nucleotides as electron donors (Hucklesby, Dalling & Hage-

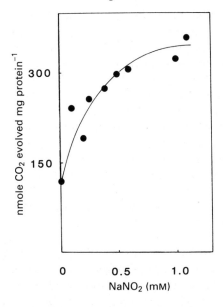

Fig. 3. Effect of nitrite concentration on the release of $^{14}CO_2$ from [1-^{14}C]-glucose 6-phosphate supplied to intact plastids for 45 minutes.

man, 1972; Nagaoka *et al.*, 1984) and require a highly electronegative electron donor such as ferredoxin. Suzuki *et al.* (1985) have demonstrated the presence in maize roots of an electron carrier and pyridine nucleotide reductase which were able to support ferredoxin-glutamate synthase activity in the presence of reduced pyridine nucleotides, but which were insufficient to support nitrite reduction. This implies that there may be other components necessary for the reduction of nitrite *in vivo*. Ninomiya & Sato (1983, 1984) have also found an electron carrier protein in plastids of non-green cultured tobacco cells with absorption and EPR spectra similar to leaf ferredoxin. The amino acid composition of leaf and radish root ferredoxin are different (Wada, Onda & Matsubara, 1986) implying that there is a distinctive molecular species of ferredoxin in non-photosynthetic tissues, which is able to act as a redox mediator in ferredoxin-dependent enzyme systems.

Current studies in our own laboratory indicate that ferredoxin and NADPH specific pyridine nucleotide reductase are induced in plastids during nitrate assimilation. Antibodies prepared against *Cucurbita pepo* leaf ferredoxin were used to detect pea root plastid ferredoxin on Western blots, and the ferredoxin was only observed in plastids taken from roots which had previously been fed nitrate. Further, pyridine nucleotide reductase activity (measured in a ferredoxin-dependent cytochrome c reduction assay) increased four to five fold over a 24 h period in plastids being fed nitrate. Two bands of NADPH-pyridine nucleotide reductase were detected after activity staining of native polyacrylamide gels for diaphorase activity, one of these bands having a very low amount of activity with NADH. Both bands of diaphorase activity were undetectable on polyacrylamide gels after electrophoresis of the soluble proteins of plastids taken from roots in which nitrate assimilation was absent. These data strongly imply the coordinated expression of root plastid nitrite reductase, ferredoxin and pyridine nucleotide reductase during the onset of nitrate assimilation, the latter two acting as a bridge between NADPH generated by the pentose phosphate pathway and nitrite reductase. The possibility that other components are necessary for nitrite reduction cannot be ruled out (Suzuki et al., 1985).

A point which is sometimes raised is that the transfer of electrons from NADPH to ferredoxin is electrochemically unfavourable, electrons being transferred from NADPH to a more electronegative redox agent. It can be calculated that even if 90% of the pyridine nucleotide were in the reduced state in vivo only a few per cent of the ferredoxin would be reduced and it is implied that this would be highly inefficient. This is a spurious argument since what matters is not the amount of ferredoxin reduced at a particular moment in time, but the turnover of electrons through ferredoxin. In other words, it does not matter if the reduction of ferredoxin is electrochemically unfavourable. As long as electrons are being used during nitrite reduction they will continue to be 'siphoned' through from NADPH, generated by glucose 6-phosphate oxidation, and it is this flux of electrons which is important, not the relative redox state of the ferredoxin. The position of ferredoxin in this sequence is equivalent to the top of the loop when siphoning petrol from the tank of a car, and as anyone who has ever done this will confirm, once there is sufficient 'pull' to surmount the energetic problems, the petrol flows freely.

The supply of carbon skeletons, ATP and reductant for amino acid synthesis

It is likely that the carbon skeletons for amino acid synthesis originate in the mitochondria, yet the study of the anaplerotic role of the Krebs cycle in respect to nitrogen assimilation has received little attention. Using developing castor bean endosperm, Nakayama, Fujii & Miura (1979) examined the incorporation of radioactive label from [2, 3-^{14}C]-succinate into amino acids. Fluorocitrate, an inhibitor of aconitase, inhibited ^{14}C incorporation into glutamate and glutamine whilst aminooxyacetate, an aminotransferase inhibitor, stimulated release of $^{14}CO_2$. These results are consistent with the assumption that the carbon skeletons of amino acids/ glutamine come from 2-oxoglutarate generated by the Krebs cycle. Journet & Douce (1983) have shown that when the intra-mitochondrial NADH to NAD^+ ratio is high in purified potato tuber mitochondria, then 2-oxoglutarate only is excreted from the matrix space. If the origin of NADH for nitrate reductase is mitochondrial then it is possible that such conditions may prevail *in vivo* during assimilation and would facilitate the integration of nitrate reduction with the primary synthesis of amino acids. However, such speculation lacks experimental support and *in vivo* data on the redox state of pyridine nucleotides and the concentration of metabolites in different subcellular compartments is needed.

Glutamine synthetase requires one molecule of ATP per molecule of glutamine synthesized. This could be generated oxidatively in the mitochondria or by substrate level phosphorylation in either the cytoplasm or root plastids since both contain a complete glycolytic sequence (Roberts, pers. comm.). There is no evidence for or against any of these possibilities and opinion is coloured by the uncertainty with regard to the intracellular location of glutamine synthetase.

Since ferredoxin-glutamate synthase is located in root plastids it is likely that a similar mechanism to that already proposed for the supply of electrons to nitrite reductase by the pentose phosphate pathway operates, and this is currently being investigated by the authors.

Summary

The compartmentation and regulation of nitrate assimilation in roots is complex. No consideration has been given in this chapter to the storage and release of nitrate from root vacuoles, about which we know very little and yet which may be crucial in regulating the assimilatory process. The integration of nitrate assimilation with carbohydrate oxidation in the rest of the cell potentially involves coordination of events between the cytoplasm, mitochondria and plastids. The source of NADH for nitrate reductase in roots remains contentious and may involve the mitochondrion or cytosol. There is little doubt that nitrite reduction is mediated by the operation of the pentose phosphate pathway in plastids and there is now evidence for the coordinated induction of root plastid ferredoxin and pyridine nucleotide reductase along with nitrite reductase. As for the study of the integration of amino acid synthesis with carbohydrate metabolism there is immense scope for future work. With improved cell fractionation procedures it is hoped that more of these issues will be tackled

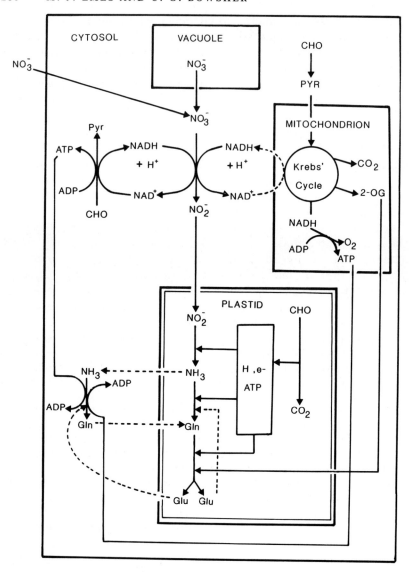

Fig. 4. Intracellular compartmentation of events involved in nitrate assimilation in higher plant roots with possible interactions between organelles and cytoplasm. 2-OG, 2-oxoglutarate; CHO, carbohydrate (G6P); Pyr, pyruvate.

in the near future. In particular, it will be necessary to develop non-aqueous fractionation procedures for root material to determine the intracellular compartmentation and concentrations of metabolites. A summary diagram is given in Fig. 4, illustrating the compartmentation of (possible) interrelated steps during nitrate assimilation in higher plant roots.

Acknowledgements

The authors are grateful to the SERC and AFRC for financial support.

References

Andrews, M. (1986). The partitioning of nitrate assimilation between root and shoot of higher plants. *Plant, Cell Env.* **9**, 511–19.

Aslam, A. & Huffaker, R. C. (1982). *In vivo* nitrate reduction in roots and shoots of barley (*Hordeum vulgare* L.) seedlings in light and darkness. *Plant Physiol.* **70**, 1009–13.

Beevers, L. & Hageman, R. H. (1983). Uptake and reduction of nitrate; bacteria and higher plants. In *Encyclopaedia of Plant Physiology*, vol. 15A, ed. A. Lauchli & R. L. Bieleski, pp. 351–97. New York: Springer-Verlag.

Blackwell, R. D., Murray, A. J. S. & Lea, P. J. (1987). Inhibition of photosynthesis in barley with decreased levels of chloroplastic glutamine synthetase activity. *J. Exp. Bot.* **38**, 1799–809.

Boland, M. J. & Benny, A. G. (1977). Enzymes of nitrogen metabolism in legume nodules. Purification and properties of NADH-dependent glutamate synthase from lupin nodules. *Eur. J. Biochem.* **79**, 355–62.

Borchert, S., Grosse, H. & Heldt, H. W. (1989). Specific transport of inorganic phosphate, glucose 6-phosphate, dihydroxyacetone phosphate and 3-phosphoglycerate into amyloplasts from pea roots. *FEBS Lett.* **253**, 183–6.

Bowsher, C. G., Hucklesby, D. P. & Emes, M. J. (1989). Nitrite reduction and carbohydrate metabolism in plastids purified from roots of *Pisum sativum* L. *Planta* **177**, 359–66.

Butt, V. S. & Beevers, H. (1961). The regulation of pathways of glucose catabolism in maize roots. *Biochem. J.* **80**, 21–7.

Calza, R., Huttner, E., Vincentz, M., Rouze, P., Galangeau, F., Vaucheret, H., Cherel, I., Meyer, C., Kronenberger, J. & Caboche, M. (1987). Cloning of DNA fragments, complementary to tobacco nitrate reductase mRNA and encoding epitopes common to nitrate reductase from higher plants. *Mol. Gen. Genet.* **209**, 552–62.

Chen, F.-L. & Cullimore, J. V. (1988). Two isoenzymes of NADH-dependent glutamate synthase in root nodules of *Phaseolus vulgaris* L.

Purification, properties and activity changes during nodule development. *Plant Physiol.* **88**, 1411–17.

Cheng, C. L., Dewdney, J., Kleinhofs, A. & Goodman, H. M. (1986). Cloning and nitrate induction of nitrate reductase mRNA. *Proc. Natn. Acad. Sci. USA* **83**, 6825–8.

Crawford, N. M., Campbell, W. H. & Davis, R. W. (1986). Nitrate reductase from squash: cDNA cloning and nitrate regulation. *Proc. Natn. Acad. Sci. USA* **83**, 8073–6.

Cullimore, J. V., Lara, M., Lea, P. J. & Miflin, B. J. (1983). Purification and properties of two forms of glutamine synthetase from the plant fraction of *Phaseolus* root nodules. *Planta* **157**, 245–53.

Cullimore, J. V. & Miflin, B. J. (1984). Immunological studies on glutamine synthetase antisera raised to the two plant forms of the enzyme from *Phaseolus* root nodules. *J. Exp. Bot.* **35**, 581–7.

Dancer, J. E. & ap Rees, T. (1989). Effects of 2, 4-dinitrophenol and anoxia on the inorganic-pyrophosphate content of the spadix of *Arum maculatum* and the root apices of *Pisum sativum*. *Planta* **178**, 421–4.

Dry, I. B., Bryce, J. H. & Wiskich, J. T. (1987). Regulation of mitochondrial respiration. In *The Biochemistry of Plants*, vol. 11, ed. D. D. Davies, pp. 213–52. London: Academic Press.

Dry, I. B., Wallace, W. & Nicholas, D. J. D. (1981). Role of ATP in nitrite reduction in roots of wheat and pea. *Planta* **152**, 234–8.

Emes, M. J., Asihara, H. & Fowler, M. W. (1979). The influence of nitrate on particulate 6-phosphogluconate dehydrogenase activity in pea roots. *FEBS Letts.* **105**, 370–2.

Emes, M. J. & England, S. (1986). Nitrogen metabolism in plastids of pea roots. In *Fundamental, Ecological and Agricultural Aspects of Nitrogen Metabolism in Higher Plants*, ed. H. Lambers, J. J. Neeteson & I. Stulen, pp. 173–6. The Netherlands: Martinus Nijhoff.

Emes, M. J. & Fowler, M. W. (1979a). Intracellular location of the enzymes of nitrate assimilation in the apices of seedling pea roots. *Planta* **144**, 249–53.

Emes, M. J. & Fowler, M. W. (1979b). The intracellular interactions between the pathways of carbohydrate oxidation and nitrate assimilation in plant roots. *Planta* **145**, 287–92.

Emes, M. J. & Fowler, M. W. (1983). The supply of reducing power for nitrite reduction in plastids of seedling pea roots (*Pisum sativum* L.). *Planta* **158**, 97–102.

Emes, M. J. & Traska, A. (1987). Uptake of inorganic phosphate by plastids purified from the roots of *Pisum sativum* L. *J. Exp. Bot.* **38**, 1781–8.

Fentem, P. A., Lea, P. J. & Stewart, G. R. (1983). Ammonia assimilation in the roots of nitrate- and ammonia-grown *Hordeum vulgare* (cv. Golden Promise). *Plant Physiol.* **71**, 496–501.

Fowler, M. W. & Sarkissian, G. S. (1975). Malic enzyme as a source of

NADPH for nitrate assimilation in roots. *Plant Sci. Letts.* **4**, 41–6.

Galangeau, F., Danel-Vedele, F., Moureaux, T., Dorbe, M.-F., Leydecker, M.-T. & Caboche, M. (1988). Expression of leaf nitrate reductase genes from tomato and tobacco in relation to light-dark regimes and nitrate supply. *Plant Physiol.* **88**, 383–8.

Gray, V. M. & Cresswell, C. F. (1984). The effects of inhibitors of photosynthetic and respiratory electron transport on nitrate reduction and nitrite accumulation in excised *Z. mays* L. leaves. *J. Exp. Bot.* **35**, 1166–76.

Hucklesby, D. P., Dalling, M. J. & Hageman, R. H. (1972). Some properties of two forms of nitrite reductase from corn (*Zea mays* L.) scutellum. *Planta* **104**, 220–33.

Jessup, W. & Fowler, M. W. (1977). Interrelationships between carbohydrate metabolism and nitrogen assimilation in cultured plant cells III. Effect of the nitrogen source on the pattern of carbohydrate oxidation in plant cells of *Acer pseudoplatanus* L. grown in culture. *Planta* **137**, 71–6.

Journet, E.-P. & Douce, R. (1983). Mechanisms of citrate oxidation by Percoll-purified mitochondria from potato tuber. *Plant Physiol.* **72**, 802–8.

Klepper, L., Flesher, D. & Hageman, R. H. (1971). Generation of reduced nicotinamide adenine dinucleotide for nitrate reduction in green leaves. *Plant Physiol.* **48**, 580–90.

Kramer, V., Lahners, K., Back, E., Privalle, L. S. & Rothstein, S. (1989). Transient accumulation of nitrite reductase mRNA in maize following the addition of nitrate. *Plant Physiol.* **90**, 1214–20.

Lambers, H., Simpson, R. J., Beilharz, V. C. & Dalling, M. J. (1982). Growth and translocation of C and N in wheat (*Triticum aestivum*) grown with a split root system. *Physiologia Pl.* **56**, 421–9.

Lightfoot, D. A., Green, N. K. & Cullimore, J. V. (1988). The chloroplast-located glutamine synthetase of *Phaseolus vulgaris* L: nucleotide sequence, expression in different organs and uptake into isolated chloroplasts. *Plant Molec. Biol.* **11**, 191–202.

Mann, A. F., Hucklesby, D. P. & Hewitt, E. J. (1978). Sources of reducing power for nitrate reduction in spinach leaves. *Planta* **140**, 261–3.

Mann, A. F., Fentem, P. A. & Stewart, G. R. (1979). Identification of two forms of glutamine synthetase in barley (*Hordeum vulgare*). *Biochem. Biophys. Res. Comm.* **88**, 515–21.

McNally, S. & Hirel, B. (1983). Glutamine synthetase isoforms in higher plants. *Physiol. Vég.* **21**, 761–79.

Miflin, B. J. (1974). The location of nitrite reductase and other enzymes related to amino acid biosynthesis in the plastids of roots and leaves. *Plant Physiol.* **54**, 550–5.

Miflin, B. J. & Lea, P. J. (1980). Ammonia assimilation. In *The Bio-*

chemistry of Plants, vol. 5, ed. B. J. Miflin, pp. 169–202. New York: Academic Press.

Nagaoka, S., Hirasawa, M., Fukishuma, K. & Tamura, G. (1984). Methyl viologen-linked nitrite reductase from bean roots. *Agric. Biol. Chem.* **48**, 1179–88.

Naik, M. S. & Nicholas, D. J. D. (1981). Relation between CO_2 evolution and *in situ* reduction of nitrate in wheat leaves. *Aust. J. Plant Physiol.* **8**, 515–24.

Naik, M. S. & Nicholas, D. J. D. (1984). Origin of NADH for nitrate reduction in wheat roots. *Plant Sci. Letts.* **35**, 91–6.

Naik, M. S. & Nicholas, D. J. D. (1986). Malate metabolism and its relation to nitrate assimilation in plants. *Phytochemistry* **25**, 571–6.

Nakayama, H., Fujii, M. & Miura, K. (1979). Glutamine synthesis in germinating castor bean endosperm. *Plant Cell Physiol.* **20**, 543–52.

Nance, J. F. (1948). The role of oxygen in nitrate assimilation by wheat roots. *Am. J. Bot.* **35**, 602–6.

Ninomiya, Y. & Sato, S. (1983). An electron carrier in nitrite reduction in proplastids of cultured tobacco cells. *Plant Cell Tissue Organ Cult.* **2**, 285–92.

Ninomiya, Y. & Sato, S. (1984). A ferredoxin-like electron carrier from non-green cultured tobacco cells. *Plant Cell Physiol.* **25**, 453–58.

Oaks, A. & Hirel, B. (1985). Nitrogen metabolism in roots. *Ann. Rev. Plant Physiol.* **36**, 345–65.

Oji, Y., Watanabe, M., Wakiuchi, N. & Okamoto, S. (1985). Nitrite reduction in barley-root plastids: dependence on NADPH coupled with glucose-6-phosphate and 6-phosphogluconate dehydrogenases, and possible involvement of an electron carrier and a diaphorase. *Planta* **165**, 85–90.

Privalle, L. S., Lahners, K. N., Mullins, M. A. & Rothstein, S. (1989). Nitrate effects on nitrate reductase activity and nitrite reductase mRNA levels in maize suspension cultures. *Plant Physiol.* **90**, 962–7.

Redinbaugh, M. G. & Campbell, W. H. (1981). Purification and characterisation of NAD(P)H: nitrate reductase and NADH: nitrate reductase from corn roots. *Plant Physiol.* **68**, 115–20.

Russell, E. W. (1973). *Soil Conditions and Plant Growth*. London: Longman.

Sarkissian, G. S. & Fowler, M. W. (1974). Interrelationship between nitrate assimilation and carbohydrate metabolism in plant roots. *Planta* **119**, 335–49.

Sawhney, S. K., Naik, M. S. & Nicholas, D. J. D. (1978). Regulation of nitrate reduction by light, ATP and mitochondrial respiration in wheat leaves. *Nature* **272**, 647–8.

Schnarrenburger, C., Oeser, A. & Tolbert, N. E. (1973). Two isoenzymes of glucose 6-phosphate dehydrogenase and 6-phosphogluconate dehydrogenase in spinach leaves. *Arch. Biochem. Biophys.* **154**, 438–48.

Simcox, P. D., Reid, E. E., Canvin, D. T. & Dennis, D. T. (1977). Enzymes of the glycolytic and pentosephosphate pathways in proplastids from the developing endosperm of *Ricinus communis* L. *Plant Physiol.* **59**, 1128–32.

Smirnoff, N. & Stewart, G. R. (1985). Nitrate assimilation and translocation by higher plants: comparative physiology and ecological consequences. *Physiologia Pl.* **64**, 133–40.

Solomonson, L. P., Howard, W. D., Yamaya, T. & Oaks, A. (1984). Mode of action of natural inactivator proteins from corn and rice on a purified assimilatory nitrate reductase. *Arch. Biochem. Biophys.* **233**, 469–74.

Suzuki, A., Gadal, P. & Oaks, A. (1981). Intracellular distribution of enzymes associated with nitrogen assimilation in roots. *Planta* **151**, 457–61.

Suzuki, A., Oaks, A., Jacquot, J.-P., Vidal, J. & Gadal, P. (1985). A natural electron donor from maize roots for glutamate synthase and nitrite reductase. *Plant Physiol.* **78**, 374–8.

Suzuki, A., Vidal, J. & Gadal, P. (1982). Glutamate synthase isoforms in rice. Immunological studies of enzymes in green leaf, etiolated leaf and root tissues. *Plant Physiol.* **70**, 827–32.

Tingey, S. V., Walker, E. L. & Coruzzi, G. M. (1987). Glutamine synthetase genes of pea encode distinct polypeptides which are differentially expressed in leaves, roots and nodules. *EMBO J.* **6**, 1–9.

Vezina, L.-P., Hope, H. J. & Joy, K. W. (1987). Isoenzymes of glutamine synthetase in roots of pea (*Pisum sativum* cv. Little Marvel) and alfalfa (*Medicago media* Pers. cv. Saranac). *Plant Physiol.* **83**, 58–62.

Vezina, L.-P. & Langlois, J. R. (1989). Tissue and cellular distribution of glutamine synthetase in roots of pea (*Pisum sativum*) seedlings. *Plant Physiol.* **90**, 1129–33.

Wada, K., Onda, M. & Matsubara, H. (1986). Ferredoxin isolated from plant non-photosynthetic tissues. Purification and characterisation. *Plant Cell Physiol.* **27**, 407–15.

Wallsgrove, R. M., Turner, J. C., Hall, N. P., Kendall, A. C. & Bright, S. W. J. (1987). Barley mutants lacking chloroplastic glutamine synthetase-biochemical and genetic analysis. *Plant Physiol.* **83**, 155–8.

Washitani, I. & Sato, S. (1977). Studies on the function of proplastids in the metabolism of in vitro-cultured tobacco cells. III. Source of reducing power for amino acid synthesis from nitrite. *Plant Cell Physiol.* **18**, 1235–41.

Willis, A. J. & Yemm, E. W. (1955). The respiration of barley plants. VIII. Nitrogen assimilation and the respiration of the root system. *New Phytol.* **54**, 163–81.

J. F. FARRAR AND J. H. H. WILLIAMS

Control of the rate of respiration in roots: compartmentation, demand and the supply of substrate

Introduction

The major source of carbon for a growing root is sucrose produced by photosynthesis in the shoot. This sucrose is used both as a source of carbon skeletons for use in biosynthesis, and as substrate in the respiratory production of the energy used to drive growth. This chapter will address two problems inherent in the flow of carbon into roots: how is the rate of respiration of roots controlled, and how is the rate of root growth adjusted to relate to the supply of sucrose available from the shoot? We will argue that the answers to these two questions are intimately related.

Unfortunately, dogma provides apparent answers for each of these questions. Those working with plant tissues at a biochemical level have long believed that respiration is controlled by adenylate supply. Those modelling plant growth, or working with whole plants, suggest that root growth is controlled by the supply of substrate from the shoot. These two views are not just contrasting, they are incompatible. It is now clear that the rate of respiration of a growing plant tissue is stoichiometrically linked to its rate of growth, and so to hold both of the above views is to believe that both adenylates and substrate simultaneously determine the flux of carbon through roots. In this chapter we will try and reconcile the sets of observations that lie behind these two hypotheses. Using roots avoids complications from photorespiration and direct use of energy from photophosphorylation that occur when considering the energy relations of green tissue.

Compartmentation and fluxes of carbohydrates in roots

If substrates play a part in regulating the rate of respiration, it is necessary to understand what determines their occurrence and concentration in the root. A reasonably complete picture of carbon flux is available for roots

167

of young barley growing quickly ($\approx$0·02 d^{-1}), based partly on compart-
mental analysis of ^{14}C kinetics (Fig. 1). The most striking feature is the
small size of the carbohydrate pool compared with the flux through it; the
total ethanol-soluble material (mostly glucose and fructose with some
sucrose), fructan and starch is 70–100 mg g^{-1} dry weight, of which the
ethanol-soluble sugars account for half (Farrar, 1985a; Bingham & Farrar, 1988; Gordon, Ryle & Powell, 1977). The flux of sucrose into the
root, averaged over 24 h, is 14 mg g^{-1} h^{-1} (Farrar, 1985a; Farrar & Jones,
1986): the ethanol-soluble sugars are thus equivalent to only 3 h of input.
Further, $\approx$60% of the soluble sugars are compartmented in vacuoles
(Farrar, 1985a; Williams and Farrar, 1990; other species are similar:
Bryce & ap Rees, 1985; Chapleo & Hall, 1989), and so metabolically
available sugar with a half-life (t$\frac{1}{2}$)$\approx$1 h is equivalent to only 1–1·5 h
import. Barley root phloem has a solute potential of about -2·53 MPa
(Warmbrodt, 1985), equivalent to roughly 800 mol m^{-3} sucrose if 80% of
solute is sucrose. As the sieve-element companion-cell complex occupies
about 0·5% of root cross-sectional area, about 20% of the 'cytosolic'
sucrose will be within phloem. Since cytosolic sugar is in rapid
equilibrium with that in the apoplast (Farrar, 1985a; Chapleo & Hall,
1989; Giaquinta et al., 1983), allowance must be made for this; overall,
cytosolic sugar will be probably about 20 mol m^{-3} (Farrar, 1985a).

If import of assimilate into the root does not equal rate of removal of
sugars from the cytosolic pool to support growth, then this pool will
potentially be subject to rapid changes in size. Export rate from barley
leaves can vary two to four-fold over a diel cycle (Farrar and Farrar,
1987) but rate of import into roots may be buffered by carbohydrate
stores in the shoot. Following selective pruning of the plant, or shoot
removal, total soluble sugars can change greatly and rapidly – at
5 mg g^{-1} h^{-1} following shoot removal, for example (Farrar, 1985a). In
spite of the potential for change, cytosolic pool size is remarkably constant both over a diel cycle (maximum change two-fold, Williams &
Farrar, unpubl.) or after selective pruning. In both cases the greater
change is seen in the vacuolar pool, which seems to buffer the cytosolic
pool to a substantial degree (Williams & Farrar, 1990). By contrast, there
is little or only a slow change in starch and fructan (Bingham & Farrar,
1988; Farrar, 1985a). Thus, whilst the maximum rates of polymer production and breakdown are 1·4–2·0 mg g^{-1} h^{-1}, vacuolar loading and unloading at up to 5 mg g^{-1} h^{-1} can occur (Fig. 1), in spite of polymer turnover
with a t$\frac{1}{2}$$\approx$15 h (Farrar, 1985a). The t$\frac{1}{2}$ for decay of respiration rate in
excised roots is 15 h (Farrar, 1981), consistent with either vacuolar or
polymeric reserves being the substrate for this respiration. The fate of
carbon leaving the cytosolic pool is, aside from temporary storage, to

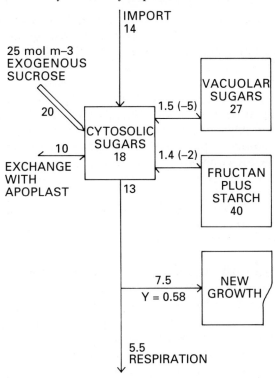

Fig. 1. Compartmentation and fluxes of carbohydrates in seminal roots of young barley plants. Numbers in boxes are pool sizes (mg g^{-1}) and by arrows are fluxes (mg g^{-1} h^{-1}). Data are derived from Farrar (1985a), Bingham & Farrar (1988), Williams & Farrar (1990) and Farrar (unpubl.).

provide both skeletons for biosynthesis of new plant structure and substrate for the respiratory processes providing energy for biosynthesis and membrane transport. The ratio between respiration and growth as a sink for carbon stays remarkably constant even when growth rate changes, at least when averaged over 24 h (Figs. 1 and 3; Farrar & Jones, 1986; Lambers, van de Werf & Konings, 1989). It is not clear how constant the flux to growth is during 24 h, but that to respiration changes only 1·5 fold over a diel cycle (Farrar, 1981). As the increased rate during the photoperiod is probably due partly to increased nutrient uptake and nitrate assimilation (see above) it is probable that growth proceeds almost constantly over 24 h, and this is supported by simple measurements of root extension rate (Farrar, unpubl.). Thus, withdrawals from the cytosolic pool will vary from about 11·5 to 14·5 mg g^{-1} h^{-1}; and the difference of 3 mg g^{-1} h^{-1} is within the buffering capacity of the vacuole.

In summary, compartmentation of sugars within the root results in a small but comparatively stable cytosolic pool. This pool will only alter appreciably in size if import into it changes for a period sustained enough to fill or deplete the vacuolar stores. The cytosolic sugar pool is thus, due to compartmentation of soluble and storage carbohydrates within the root, suitable for a regulatory role. Since uptake into the cytosol can proceed twice as quickly as loading of the vacuole (Fig. 1) experimental increases of its size are easily achieved.

Control of respiration rate by substrates

There are many reports that carbohydrate content of a tissue is correlated with its respiration rate (Fig. 2; Penning de Vries, Witlage & Kremer,

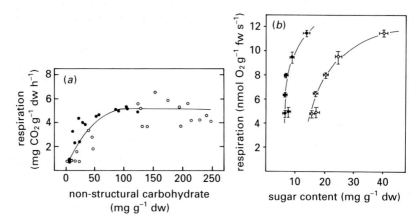

Fig. 2. Relationships between respiration rate and carbohydrate content of roots. (*a*) Respiration and non-structural carbohydrate in wheat plants (Penning de Vries *et al.*, 1979). (*b*) Respiration and cytosolic (solid circles) and total (open circles) ethanol-soluble sugars in barley roots (Williams & Farrar, 1990).

1979; and references in Farrar, 1985b). The relationship is frequently hyperbolic (Lambers, 1985). The simplest interpretation of this correlation is that it is a causal one – that substrate limits respiration so that when carbohydrate content is increased, respiration rises in consequence. It has even been suggested that at high sugar content, the activity of the alternative oxidase is increased to consume the excess sugar in the tissue (Lambers, 1982).

The commonplace observations that either raised light intensity or raised partial pressure of CO_2 increase plant growth are also most easily interpreted as showing that when substrate supply is enhanced via increased photosynthesis, then growth and thus respiration (see below) are increased. The mechanism for such a simple substrate-regulated respiration has rarely been suggested, it being seemingly implicit that it occurs by mass action (Day & Lambers, 1983) perhaps accompanied by rather loose control of glycolysis. Certainly, models of plant growth that assume that flux to structural material or to respiration is proportional to carbohydrate pool size have been very successful (Reynolds & Thornley, 1982; Johnson, 1985; Farrar, 1990).

Two cautionary notes must be sounded. First, it is not clear why total carbohydrate in a tissue should correlate with the rate of any metabolic process; only cytosolic sugars would be expected to show a correlation. There is a correlation between cytosolic sugars and respiration rate in barley roots (Fig. 2) but this may not be a general phenomenon. Second, one simple consequence of the hypothesis of substrate limitation is that the respiration of a tissue should rise if extra substrate is supplied to it; this is not usually the case, even when the tissue in question would be expected to be substrate deficient, such as roots from plants darkened for 24 h (Farrar, 1981; Bryce & ap Rees, 1985; Bingham & Farrar, 1988; Crawford & Huxter, 1977). In passing it may be noted that when sugars are supplied exogenously at a concentration sufficient to increase root carbohydrate content markedly and rapidly (say 20 mol m^{-3} for barley roots), any stimulation of respiration due to the energy cost of taking up the sugar would be very small, in the order of 2% of control respiration rate in the case of barley roots if the stoichiometry of uptake is 1 mol ATP mol^{-1} sucrose. One situation where supplying sugar does increase respiration rate is in roots recently severed from the shoot system; the fall in respiration that follows severence can be restored to control rates by adding sucrose (Saglio & Pradet, 1980; Williams & Farrar, 1990). As will be seen below, this effect is transient.

Control of respiration by adenylates

The most straightforward evidence that respiration in roots is under the control of adenylates results from adding uncouplers to them. Almost without exception, an increase in oxygen uptake is found. The explanation for this is that the uncouplers, such as DNP, carbonylcyanide chloromethoxyphenyl hydrazone (CCCP), and carbonylcyanide p-

Table 1. *Components of respiratory oxygen uptake by intact plant tissues. Parameters of the equation $v_t=\varrho V_{alt}+\phi V_{cyt}+v_{res}$ are tabulated, with measured respiration (v_t) at 100%. V_{alt} and V_{cyt} are the capacities of the alternative and cytochrome pathways, respectively, and ϱ and ϕ the degree of engagement (over the range 0–1) of those pathways*

Species	Plant part	ϱ	V_{alt}	ϕ	V_{cyt}	Ref.
Phaseolus vulgaris	root	0·07	68	1·00	89	2
Spinacea oleracea	root	0·65	40	0·77	96	2
Pisum sativum	root	0·25	44	1·00	80	6
Triticum vulgare	root	0·88	22	0·78	72	2
Hordeum distichum	root	0·07	33	0·91	85	4
Hordeum distichum on 2% nutrient	root	0·07	77	≤0·57	≥160	5
Hordeum distichum leaf pruned	root	0	59	0·88	103	4
Hordeum distichum root pruned	root	0·09	23	0·91	88	4
Lolium perenne	root	0·75	50	0·61	90	3
Zea mays	root	0·66	29	0·59	86	2
Phaseolus vulgaris	leaves	0·10	93	0·81	92	1
Fatsia japonica	leaves	0·25	78	0·99	77	7
Hordeum distichum	leaves	0·53	29	0·66	83	8
Lolium perenne	leaves	<0·86	88	<0·54	160	3

References: 1, Azcon-Bieto, Lambers & Day, 1983; 2, Day & Lambers, 1983; 3, Day *et al.*, 1986; 4, Bingham & Farrar, 1988; 5, Bingham & Farrar, 1988; 6, de Visser & Blacquiere, 1986; 7, Burgos, Araus & Azcón-Bieto, 1987; 8, D. Ll. Lewis & Farrar (unpubl.).

trifluoromethoxyphenyl hydrazone (FCCP), are protonophores and thus permit recycling of protons through the mitochrondrial inner membrane in the absence of ATP synthesis. A faster rate of electron flow through the mitochondrial electron transport chain exceeds the adenylate-limited rate than that obtained before adding uncoupler (Beevers, 1961; Bingham & Farrar, 1988; Day & Lambers, 1983). Some examples of uncoupling are given in Table 1.

This is not to say that the cytochrome path is adenylate-limited. The mitochondrial electron transport chain (METC) has a branch point at ubiquinone, from which electrons can enter either the cytochrome path or the non-phosphorylating alternative path. Total respiratory oxygen uptake v_t can be described by:

$$v_t = \phi V_{cyt} + \varrho V_{alt} + v_{res}$$

where V_{cyt} is the capacity of the cytochrome path, V_{alt} the capacity of the alternative path, ϕ and ϱ the degree of engagement of the cytochrome and alternative paths respectively, and v_{res} is residual respiration (that not inhibited by cyanide and salicylhydroxamate (SHAM) together). The parameters of this equation can be estimated by measuring rates of tissue respiration in increasing concentrations of SHAM in the presence and absence of cyanide, and drawing a ϱ or Bahr-Bonner plot of the resulting data (Bahr & Bonner, 1973; Theologis and Laties, 1978). To investigate the capacity of the cytochrome path, and see if it exceeds its activity *in vivo*, it is necessary to allow for the contribution of the alternative path to electron flow. An inhibitor of the alternative path, such as SHAM, can be used with intact tissues if sufficient care is taken (Bingham & Farrar, 1987; Möller *et al.*, 1988). SHAM inhibits much of the uncoupler-induced rise in respiration of a tissue, showing that relief of adenylate limitation on respiration results in an increased flow down the alternative path as well as the cytochrome path. Thus, adenylate control of mitochondrial electron transport is likely to be exercised largely at complex I (though complexes III and IV will also be involved). Table 1 shows the relative capacities and activities of the two pathways in a variety of tissues.

Judicious use of uncouplers therefore suggests that adenylates limit the rate of respiration in root tissues. The second line of evidence suggesting adenylate control is that an increase in the demand for ATP results in an increased rate of respiration. The classical example is salt respiration, the rise in respiration rate shortly after adding salts (Briggs, Hope & Robertson, 1961; Bingham & Farrar, 1989; Lüttge, Cram & Laties, 1971; Milthorpe & Robertson, 1948; Willis & Yemm, 1955). The stoichiometry between net ion uptake and increased respiration strongly suggests that increased use of ATP to drive active uptake relieves the adenylate limitation on respiration. Further evidence of adenylate limitation is provided by the existence of the Pasteur effect (Beevers, 1961).

It is commonly observed that rate of root respiration falls following the darkening of the shoot, and rises following shoot re-illumination, sometimes within 12 min (Farrar, 1981; Hansen, 1980; Neales & Davies, 1966; Osman, 1971). Although one explanation is that supply of substrate from the shoot limits respiration, it is more likely that some energy-utilizing process, such as uptake of nutrients delivered to the root surface in the transpiration stream, is responsible for the changes in respiration via demand for adenylates (Farrar, 1981; Massimino *et al.*, 1980; Deane-Drummond, Clarkson & Johnson, 1980; Hansen, 1980).

In summary, there is excellent reason to believe that the rate of root respiration is under the immediate control of adenylates. This control will be exercised primarily in the mitochondria on electron transport and NADH oxidation, whilst cytosolic ADP/ATP ratio may control flux through glycolysis via its effect on PFK so that glycolytic carbon flux matches mitochondrial electron transport. Whether fructose 2, 6-bisphosphate is involved in this control is, as yet, uncertain.

The functions of root respiration

If adenylate turnover does indeed limit respiration, then control of respiration rate will ultimately be exercised by the rate-control of energy consuming processes in the root. Here we briefly survey these and their quantitative importance. In passing, it should be noted that even if a seminal root apex is elongating at its maximum possible rate, this cannot be so for the root system as a whole, as not only is increased branching possible but also first-order branches frequently grow well below the rate at which they are capable (Farrar & Jones, 1986).

The relationship of outstanding importance is that between respiration and growth (Fig. 3). The two are linearly related, the slope of the line being a yield coefficient describing the efficiency with which each unit of substrate is converted into new plant material. The low rate of respiration at zero growth rate is sometimes interpreted as the respiratory cost of maintaining plant material. A useful consequence of the relationship between rates of growth and respiration is that the latter can be used as a non-destructive assay of the former; a profound one, that to understand the control of respiration is to understand the control of dry matter increment.

It is possible to estimate the proportion of respiration that is concerned with growth, with ion uptake, or with maintenance at 25–40, 10–50 and 12–50% respectively (Lambers *et al.*, 1989), these values changing with species and ontogeny. Such estimates, arrived at by regression analysis, are limited in that, for example, both growth and maintenance include a component for protein synthesis, and assimilation rather than just uptake of NO_3^- will be included in ion uptake. Estimates can also be made by calculating the rate of respiration necessary to provide energy for a process occurring at a known rate in the root. Such estimates (Table 2) are of necessity approximate, yet are useful both because they give a check on values derived from regression analysis, and indicate in more detail which processes are expensive in terms of respiratory energy. The figures in

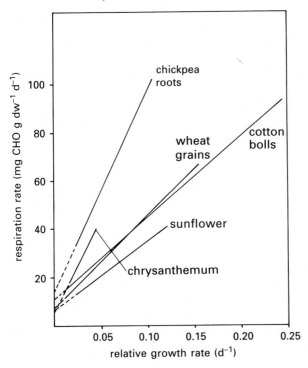

Fig. 3. Relationship between respiration rate and relative growth rate in plants. Data refer to whole plants unless stated and are recalculated where necessary from Kallarackal & Milburn, 1985 (chickpea); Hughes, 1973 (chrysanthemum); Lambers, 1984 (wheat grains); Thornley & Hesketh, 1972 (cotton bolls); and Kidd, West & Briggs, 1921 (sunflower). A similar relationship has been shown for barley roots between respiration and nitrogen increment (Folkes & Yemm, 1958).

Table 2 are certainly within the ranges given above; they also suggest, strikingly, that the proportion of NO_3^- assimilation that occurs in the root will have a dramatic effect on respiration rate. Estimates for cereals range from 3 (Minotti & Jackson, 1970) to 75% (Rufty, Jackson & Raper, 1981; Murray & Ayres, 1986) of NO_3^- being root-reduced; these fractions would account for 10–240% of barley root respiration. Either the literature estimates of the cost of NO_3^- reduction are too high, or only $\approx 10\%$ of NO_3^- is root-reduced; certainly the importance of N metabolism for determining respiration rate has been underestimated.

Table 2. *Estimated respiratory costs of major processes in roots of young barley plants. These costs (mg CHO g dw^{-1} h^{-1}) are approximate and calculated from the following assumptions. Relative growth rate 0·2 d^{-1}; cost of dry matter increment 0·23 C respired (C incorporated)$^{-1}$; turnover of dry matter ($t_{1/2}$) 15 d; carbohydrate storage costs 0·05 mol mol^{-1}; nutrient content 8% dry mass, of which 50% needs 1 mol ATP mol^{-1} to enter the root; N is 4% dry mass, assimilation costs 1 mol C mol N^{-1}, and 10% is assimilated in the roots; cations are at 300 mol m^{-3} in the cytosol with $t_{1/2}=0·8$ h and reabsorption costs 1 mol ATP mol^{-1}; oxidation of 1 mol hexose yields 30 mol ATP; root:shoot ratio=1:3*

Net synthesis of structural material (protein, lipid, wall)	1·83	27%	'growth', 70%
Net cation uptake	0·72	net ion uptake,	
Net NO$_3^-$ assimilation	2·10	43%	
Maintenance of ionic status ('pump and leak')	0·56	'maintenance', 30%	
Turnover of structural material	0·56		
Turnover of non-structural carbohydrate	0·85		
	6·62		

Sucrose supply and the capacity for respiration

Changing the source:sink ratio of barley plants by selective pruning has a profound effect on both carbohydrate content and respiration rate of the root system. A week after pruning young plants to either the first leaf blade or to one seminal root, a two-fold difference in growth rate (due largely to differential growth of lateral roots), soluble sugar and respiration rate, occurs (Farrar & Jones, 1986). Changes nearly as great can be found 16–24 h after pruning (Bingham & Farrar, 1988; Williams & Farrar, 1990) and it is the response of roots to treatments of this duration that will be considered further.

The changes in respiration do not involve engagement of the alternative oxidase, even when carbohydrate status is high (Bingham & Farrar, 1988), thus lending no support to the idea of the alternative path as an energy overflow (Lambers, 1985). The value of ϱ (the proportional engagement) is always <0·1 and capacity does not change. The primary response is confined to the cytochrome pathway (Bingham & Farrar, 1988). Nor does sucrose, added to the roots 10 min before measuring

Table 3. *Effects of selective pruning of barley plants on the respiration rate of roots (nmol O_2 g fw^{-1} s^{-1}): plants were pruned 20 h before measurement*

	Control	Shoot-pruned to one leaf	Root pruned to one seminal
Respiration rate	3·5±0·2	1·8±0·1	4·8±0·4
Increase with 25 mol m^{-3} sucrose	0·2±0·1	0·1±0·2	0·2±0·2
Increase with FCCP	1·2±0·1	0·7±0·1	1·2±0·1
Increase with FCCP, SHAM and sucrose	0·3±0·1	0·3±0·1	0·4±0·2
V_{cyt}	3·4±0·4	1·9±0·1	4·3±0·3
% engagement of cytochrome path	91	88	91

(From Bingham & Farrar, 1988.)

0respiration rate, cause a stimulation of respiration, even though it can increase cytosolic pool size by ≈25% in this time. At all source:sink ratios, the mitochondrial path appears to be operating at nearly full capacity; in the presence of the uncoupler FCCP, sucrose as substrate, and SHAM to inhibit the alternative pathway, electron flux down the cytochrome chain can be increased by only ≈10% (Tables 1, 3). (As respiratory oxygen consumption in the absence of SHAM is substantially increased by FCCP, both mitochondrial complex 1 and glycolytic and TCA cycle enzymes must be present in excess relative to the cytochrome path.) Thus, the alterations in respiration that follow within 24 h of changing root:shoot ratio are due to altered capacity of the cytochrome path. As root pruning increases, and shoot pruning decreases, respiration it appears that cytochrome path capacity can increase or decrease within 24 h (Table 4). This conclusion rests on the assumptions that the inhibitors penetrate fully into intact tissues, and that carbon flux from sucrose to the mitochondria is not limiting respiration rate; the latter is unlikely as FCCP causes ≈30% increase in oxygen uptake (Bingham & Farrar, 1988).

Changes in cytochrome path capacity can also be brought about by sucrose alone. Following partial shoot removal, the reduction in respiration rate can be prevented by supplying sucrose (25 mol m^{-3}) either immediately or after 24 h. Sucrose supply results in no reduction, or a rise to control rates, respectively (Table 4). In each case the capacity of the

Table 4. *Control of capacity for respiratory oxygen uptake by sucrose.*
Young barley plants were selectively pruned and, either immediately or
24 h later, their root systems were supplied with 25 mol m^{-3} sucrose for
24 h; root systems were then excised and oxygen uptake measured by
oxygen electrode. Data are presented as percent of the control (non-
pruned) plants not supplied with sucrose

	Measured 24 h after pruning		Measured 48 h after pruning
	−sucrose	+sucrose	+sucrose
(a) respiration rate			
Control	100	125	124
All leaves removed	46	85	74
Shoot shaded	44	94	84
(b) capacity of cytochrome path			
Control	100	147	—
Shoot pruned to one leaf	66	27	109
Root pruned to one seminal	160	—	—

From Bingham & Farrar (1988) and Williams & Farrar (1990). Coefficient of
variation≈15%; −, not determined.

cytochrome path was responsible for the change in rate (Bingham &
Farrar, 1988). The obvious conclusion is that sucrose is the agent responsible for the source-induced changes in sink respiratory capacity.

After selective pruning, respiration rate and carbohydrate content
change at different rates. The one remaining seminal on a root-pruned
plant shows a large and sustained increase in soluble carbohydrate within
1 h, but respiration does not change within 8 h; conversely pruning or
shading the shoot results in a fall in carbohydrate but not respiration
within 1 h (Fig. 4). These and other differences of detail in the kinetics of
change following pruning make it most unlikely that carbohydrate
exercises a simple control over respiration rate. However, a clear effect of
substrate can be seen: when root respiration is reduced following shoot
pruning, addition of sucrose to the root can indeed cause an increase in
rate back to control values, as can glucose addition to maize root tips
respiring at a reduced rate after excision (Saglio & Pradet, 1980). This
sensitivity to substrate lasts only for a short time – about 7 h after pruning; at longer periods, it is quite without effect (Fig. 4). A fundamental

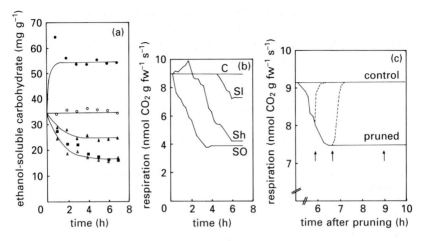

Fig. 4. Effect of selective pruning of barley plants on (*a*) the ethanol-soluble carbohydrate content, and (*b*) the respiration rate of their roots. Key: ○ or C, control; ● or SI, pruned to one seminal root; ▲, pruned to one leaf; ■ or Sh, shaded; △ or SO, pruned to no leaves. (*c*) Effect on root respiration of adding 25 mol m^{-3} sucrose 5·8, 6·6 and 9 h after pruning the shoot to one leaf. Solid lines, respiration of control and pruned plants with no sucrose additions. Dashed lines, respiration of roots on shoot-pruned plants after addition of sucrose. Respiration measured non-destructively and continuously, as CO_2 evolution by IRGA. From Williams & Farrar (1990).

change in sensitivity to substrate thus occurs at about the time that cytochrome path capacity may begin to change. The altered sensitivity to sucrose is not a reflection of a change in compartmentation within the root. For example, following pruning to one leaf, the cytosolic and vacuolar sugars of the root both fall to about 75% of control values (Williams & Farrar, 1990).

Thus, sucrose can raise respiration rate by acting as a substrate only under a restricted set of conditions: for about 7 h after the root has been depleted of sugar by pruning. At longer times, or at the end of a dark period in an intact plant, it has no direct effect. But it does have a long term effect in raising root respiration, via the capacity of the cytochrome pathway, when supplied for 24 h. Both glucose and fructose separately share this ability, but osmotica such as mannitol and polyethylene glycol (PEG) 10 000 do not (Williams & Farrar, 1990). To summarize: it appears that sucrose sets the capacity (by coarse control) and adenylates the degree of expression of that capacity (by fine control) of mitochondrial respiration.

Evidence supporting a non-nutritional role for sucrose

The most persuasive and elegant evidence of sucrose providing coarse control comes from Douce's laboratory (Douce *et al.* this volume). Sycamore cells in suspension culture show loss of mitochrondrial function – probably due to loss of individual mitochondria – when starved of sucrose, and the beginning of the decline in mitochondrial function coincides with the fall in endogenous sucrose concentration.

Other work with tissue culture suggests that when callus cultures are limited by a nutrient other than sucrose, the response to sucrose does not follow the pattern suggested above (Klerk-Kliebert & van der Plas 1985).

There is evidence from roots supporting a non-nutritional role of sucrose. When cultured aseptically, tomato roots can be induced to show a hyperbolic relationship between concentration of sucrose supplied exogenously and respiration rate; this was achieved by sugar deprivation for 48 h followed by 68 h on the chosen concentration of sugar (Morgan & Street, 1959) and this response is paralleled by sucrose-controlled growth rate (Dawson & Street, 1959). These are just those findings that would be expected if sucrose worked via coarse rather than fine control. Similarly, excised pea root sections show a sucrose – induced rise in both respiration and extension growth 12–24 h after sucrose presentation (Audus & Garrard, 1953).

From time to time sucrose has been implicated in the control of a number of morphogenetic processes, including differentiation of phloem (Wetmore & Rier, 1963; Jeffs & Northcote, 1967; Wright & Northcote, 1972; Aloni, 1987), floral evocation (Bernier, 1988) and the initiation of root meristems (Street, 1969) and roots in callus (Wright & Northcote, 1972). The latter is particularly significant for the present argument as many of the effects we have observed would be produced by initiation of lateral roots alone. Certainly cell division in root meristems is dependent on sucrose supply (van't Hoff, Hoppin & Yagi, 1973). Far better and more systematic evidence is needed to support the involvement of sucrose in such morphogenetic processes. Sucrose may also be responsible for endogenous control of photosynthesis, possibly again involving coarse control (Williams & Farrar, 1988) and indeed sucrose has been shown to suppress chlorophyll synthesis and reduce chloroplast number in carrot callus culture (Edelman & Hanson, 1971).

In other taxa, sugars have long been accepted as playing the type of role we suggest for sucrose in higher plants; thus glucose regulates respiration via coarse control in yeasts (Käpelli, 1986) and both catabolite repression and sugar-inducible enzymes are widely reported from bacteria.

Just how sucrose might bring about these, and other, changes is the subject of the next section.

Mechanism of coarse control by sucrose

It is well to be clear at the outset: we are a long way from understanding how sucrose exercises coarse control. There may be more than one mechanism, since both central metabolism and morphogenesis can be affected; there must be effects on the demand for respiratory energy as well as the METC, or fine control by adenylates would simply suppress the actual rate of respiration in spite of an increase in capacity.

Starting with mitochondria, a basic question is whether the number or the METC of each mitochondrion is altered. This is a particularly difficult question to answer in such a heterogeneous tissue as a growing root where there may be discrete populations of mitochondria (there is no reason to believe those in meristems to be identical to those in mature cells). Even within cells there could be a hierarchy of mitochondria, perhaps age-dependent, which respond differentially to treatments. In the relative homogeneity of tissue culture, mitochondrial number per cell changes, and additionally there are modifications of the remaining mitochondria (Douce *et al.*, this volume). Whatever the nature of mitochondrial modification it is likely to involve both the nuclear and mitochondrial genome, which cooperate in specifying mitochondrial structure (Hawkesford & Leaver, 1987).

That sugars can indeed regulate the synthesis of individual proteins in higher plants is becoming established from non-mitochondrial systems. Sucrose-sucrose fructosyl transferase is synthesized in leaf tissue of high sucrose status (Cairns & Pollock, 1988; Wagner, Weimken & Matile, 1986). The fructan hydrolase, phleinase, increases in activity at low glucose concentrations due to *de novo* synthesis (Yamomoto & Mino, 1987, 1989). Sucrose synthase activity increases in detached leaves of eggplant when their sucrose content is increased (Claussen, Loveys & Hawker, 1986). Invertase, both acid and neutral, increases in activity with increasing glucose concentration in *Saccharum* cell cultures (Maretzki, Thom & Nickell, 1974), and this increase is sensitive to inhibitors of protein synthesis. Similarly, invertase synthesis in internodes of *Saccharum* is partly controlled by sugars (Gayler & Glasziou, 1972). Alkaline invertase is induced by sucrose in sugar beet suspension cultures (Masuda, Takahashi & Sugawara, 1988). A cycloheximide-sensitive increase in invertase is found 5–10 h after sucrose, glucose or fructose presentation to *Avena* stem segments (Kaufman *et al.*, 1973).

It is not always carbohydrate metabolism that responds to sugar status.

There is a suggestion that nitrate reductase is synthesized partly in response to glucose supply (Hänisch ten Cate & Breteler, 1981). As noted above, effects on energy-demanding processes are attractive as a means of relieving adenylate limitation and giving purpose to the increased capacity for respiration. Other clear effects of sugars cannot be ascribed to any single area of metabolism. Baysdorfer & van der Woude (1988) have shown that newly-synthesized proteins in maize roots differ following sugar starvation or feeding with very high concentrations of sugar, a few proteins showing either more or less synthesis under these treatments; pea root meristems alter patterns of protein synthesis when starved of carbohydrate (Webster, 1980). Our own gels of protein extracted from barley roots, which have been subject to shoot pruning with or without sucrose feeding, show that numerous bands show changes with treatment in the intensity of labelling by 35S-methionine but we cannot yet assign functions to any of the proteins so affected. We can show an increase in acid invertase activity, and in total protein, at high sugar status.

It may not be the sucrose molecule itself that is recognized, although sucrose may be unique in promoting the growth of dicot roots (Street, 1969). In barley roots, exogenous glucose is as effective as sucrose in coarse control (Williams & Farrer, unpubl.); as these roots contain far more hexose than sucrose, it may be another sugar that exercises control. It may not even be cytosolic concentration of a sugar, rather flux through a particular part of the respiratory system, that is effective.

Overall it may be suggested that sucrose or some product of its metabolism can control gene expression, probably increasing the transcription of selected genes. A mechanism is needed that increases both METC capacity and the demand for adenylates. One hypothesis is that sucrose (or some other sugar) can regulate, via transcription, one or more processes consuming respiratory energy, but not METC directly; this increased capacity for ATP consumption increases the flux through the METC, increasing the reduction of its components. Just as protein phosphorylation is controlled by the degree of reduction of plastoquinone in photosynthetic electron transport, the reduced METC component might then trigger, indirectly, the production of more METC components.

Conclusions

In addition to acting as a substrate, sucrose can exercise coarse control over respiratory metabolism. Far more evidence is needed for all stages of the arguments forwarded here. Nevertheless the prospect is clear: sucrose, the major end-product of photosynthesis in source leaves and the major substrate for metabolism in sinks, can act to regulate the capacity of sink metabolism to match the availability of assimilate. This conclusion has profound consequences for understanding the coordinated development of different parts of a growing plant, for the concepts of source and sink limitation of growth, and for the control of partitioning in plants.

Acknowledgements

We would like to thank the AFRC and MAFF for financial support. We are indebted to Ian Bingham, Elizabeth McDonnell-Zaba and Mike Williams for valuable advice and discussion.

References

Aloni, R. (1987). Differentiation of vascular tissue. *Ann. Rev. Plant Physiol.* **38**, 179–204.

Audus, L. J. & Garrard, A. (1953). Studies on the growth and respiration of roots. 1. The effects of stimulatory and inhibitory concentrations of β-indolacetic acid on root sections of *Pisum sativum. J. Exp. Bot.* **4**, 330–48.

Azcón-Bieto, J., Lambers, H. & Day, D. A. (1983). Effect of photosynthesis and carbohydrate status on respiratory rates and the involvement of the alternative path in leaf respiration. *Plant Physiol.* **72**, 598–600.

Bahr, J. T. & Bonner, W. D. (1973). Cyanide-insensitive respiration. II. Control of the alternate pathway. *J. Biol. Chem.* **248**, 3446–50.

Baysdorfer, C. & van der Woude, W. J. (1988). Carbohydrate responsive proteins in the roots of *Pennisetum americanum. Plant Physiol.* **87**, 566–70.

Beevers, H. (1961). *Respiratory Metabolism in Plants.* Evanston: Row-Peterson.

Bernier, G. (1988). The control of floral evocation and morphogenesis. *Ann. Rev. Plant Physiol. Plant Molec. Biol.* **39**, 175–219.

Bingham, I. J. & Farrar, J. F. (1987). Respiration of barley roots: assessment of activity of the alternative path using SHAM. *Physiologia Pl.* **70**, 491–8.

Bingham, I. J. & Farrar, J. F. (1988). Regulation of respiration in roots of barley. *Physiologia Pl.* **73**, 278–85.

Bingham, I. J. & Farrar, J. F. (1989). Activity and capacity of respiratory pathways in barley roots deprived of inorganic nutrients. *Plant Physiol. Biochem.* **27**, 847–54.

Briggs, G. E., Hope, A. B. & Robertson, R. N. (1961). *Electrolytes and Plant Cells.* Oxford: Blackwell.

Bryce, J. H. & ap Rees, T. (1985). Effects of sucrose on the rate of respiration of the roots of *Pisum sativum. J. Plant Physiol.* **120**, 363–7.

Burgos, G., Araus, J. L. & Azcón-Bieto, J. (1987). The effect of temperature on respiratory pathways of *Fatsia japonica* leaves. In *Plant Mitochondria*, ed. A. L. Moore & R. B. Beechey, pp. 389–92. New York: Plenum Press.

Cairns, A. J. & Pollock, C. J. (1988). Fructan biosynthesis in excised leaves of *Lolium temulentum* L. 2. Changes in fructosyl transferase activity following excision and application of inhibitors of gene expression. *New Phytol.* **109**, 407–13.

Chapleo, S. & Hall, J. L. (1989). Sugar unloading in roots of *Riccinus communis* L. II. Characteristics of the extracellular apoplast. *New Phytol.* **111**, 381–90.

Claussen, W., Loveys, B. R. & Hawker, J. S. (1986). Influence of sucrose and hormones on the activity of sucrose synthase and invertase in detached leaves and leaf sections of eggplants (*Solanum melongena*). *J. Plant Physiol.* **124**, 345–57.

Crawford, R. M. M. & Huxter, T. J. (1977). Root growth and carbohydrate metabolism at low temperatures. *J. Exp. Bot.* **28**, 917–25.

Dawson, J. R. O. & Street, H. E. (1959). Growth responses of excised roots of red clover. *Bot. Gaz.* **120**, 227–34.

Day, D. A. & Lambers, H. (1983). The regulation of glycolysis and electron transport in roots. *Physiologia Pl.* **58**, 155–60.

Day, D. A., Vos, O. C. de, Wilson, D. & Lambers, H. (1986). Regulation of respiration in the leaves and roots of two *Lolium perenne* populations with contrasting mature leaf respiration rates and crop yields. *Plant Physiol.* **78**, 678–83.

Deane-Drummond, C. E., Clarkson, D. T. & Johnson, C. B. (1980). The effect of differential root and shoot temperature on the nitrate reductase activity, assayed in vivo and in vitro in roots of *Hordeum vulgare* (barley). *Planta* **148**, 455–61.

Edelman, J. & Hanson, A. D. (1971). Sucrose suppression of chlorophyll synthesis in carrot callus cultures. *Planta* **98**, 150–6.

Farrar, J. F. (1981). Respiration rate of barley roots: its relation to growth, substrate supply and the illumination of the shoot. *Ann. Bot.* **48**, 53–63.

Farrar, J. F. (1985a). Fluxes of carbon in roots of barley plants. *New Phytol.* **99**, 57–69.

Farrar, J. F. (1985b). The respiratory source of CO_2. *Plant, Cell Env.* **8**, 427–38.

Farrar, J. F. (1990). The carbon balance of slow-growing and fast-growing species. In *Variations in Growth Rate and Productivity*, ed. H. Lambers, pp. 241–56. The Hague: SPB.

Farrar, J. F. & Jones, C. L. (1986). Modification of respiration and carbohydrate status of barley roots by selective pruning. *New Phytol.* **102**, 513–21.

Farrar, S. C. & Farrar, J. F. (1987). Effects of photon fluence rate on carbon partitioning in barley source leaves. *Plant Physiol. Biochem.* **25**, 541–8.

Folkes, B. F. & Yemm, E. W. (1958). The respiratory of barley plants. X. Respiration and metabolism of amino-acids in germinating grain. *New Phytol.* **57**, 106–31.

Gayler, K. R. & Glasziou, K. T. (1972). Regulatory roles for acid and neutral invertases in growth and sugar storage in sugar cane. *Physiologia Pl.* **27**, 25–31.

Giaquinta, R. T., Lin, W., Sadler, N. L. & Franceschi, V. R. (1983). Pathway of phloem unloading of sucrose in corn roots. *Plant Physiol.* **72**, 362–7.

Gordon, A. J., Ryle, G. J. A. & Powell, C. E. (1977). The strategy of carbon utilization in uniculm barley. 1. The chemical fate of photosynthetically assimilated ^{14}C. *J. Exp. Bot.* **28**, 1258–69.

Hänisch ten Cate, C. H. & Breteler, H. (1981). Role of sugars in nitrate utilization by roots of dwarf bean. *Physiologia Pl.* **52**, 129–35.

Hansen, G. K. (1980). Diurnal variation of root respiration rates and nitrate uptake as influenced by nitrogen supply. *Physiologia Pl.* **48**, 421–7.

Hawkesford, M. J. & Leaver, C. J. (1987). Structure and biogenesis of the plant mitochondrial inner membrane. In *Plant Mitochondria*, ed. A. L. Moore & R. B. Beechey, pp. 251–63. New York: Plenum Press.

Hughes, A. J. P. (1973). A comparison of the effects of light intensity and duration on *Chrysanthemum morifolium* in controlled environments. *Ann. Bot.* **37**, 275–85.

Jeffs, R. A. & Northcote, D. H. (1967). The influence of indol-3yl acetic acid and sugar on the pattern of induced differentiation in plant tissue culture. *J. Cell. Sci.* **2**, 77–88.

Johnson, I. R. (1985). A model of the partitioning of growth between the shoots and roots of vegetative plants. *Ann. Bot.* **55**, 421–31.

Kallarackal, J. & Milburn, J. A. (1985). Respiration and phloem translocation in the roots of chickpea (*Cicer arietinum*). *Ann. Bot.* **56**, 211–18.

Käppeli, O. (1986). Regulation of carbon metabolism in *Saccharomyces cerevisiae* and related yeasts. *Adv. Microbiol. Res.* **28**, 181–209.

Kaufman, P. B., Ghosheh, N. S., Lacroix, J. D., Soni, S. L. & Ikuma, H. (1973). Regulation of invertase levels in *Avena* stem segments by gibberellic acid, sucrose, glucose, and fructose. *Plant Physiol.* **52**, 221–8.

Kidd, F., West, C. & Briggs, G. E. (1921). A quantitative analysis of the growth of *Helianthus annuus*. *Proc. Roy. Soc. London* **B92**, 368–84.

Klerk-Kliebert, de Y. M. & van der Plas, L. H. W. (1985). Growth and respiration of soybean cells growth in batch suspension culture with various glucose concentrations. *Plant Cell Tissue Cult.* **4**, 225–33.

Lambers, H. (1982). Cyanide-resistant respiration: a nonphosphorylating electron transport pathway acting as an energy overflow. *Physiologia Pl.* **55**, 478–85.

Lambers, H. (1984). Respiratory metabolism in wheat. In *Wheat Growth and Modelling*, eds. W. Day & R. K. Atkin, pp. 123–7. New York: Plenum Press.

Lambers, H. (1985). Respiration in intact plants and tissues: Its regulation and dependence on environmental factors, metabolism and invaded organisms. In *Encyclopedia of Plant Physiology*, vol. 18, ed. R. Douce & D. A. Day, pp. 418–73. Berlin: Springer-Verlag.

Lambers, H., van der Werf, A. & Konings, H. (1989). Respiratory patterns in roots in relation to their functioning. In *Plant Roots, the Hidden Half*, ed. Y. Waisel, A. Eshal & V. Kafkafi. (In press.)

Lüttge, U., Cram, W. J. & Laties, G. G. (1971). The relationship of salt stimulated respiration to localised ion transport in carrot tissue. *Z. Pflphysiol.* **64**, 418–26.

Maretzki, A., Thom, M. & Nickell, L. G. (1974). Utilisation and metabolism of carbohydrates in cell and callus cultures. In *Tissue Culture and Plant Science 1974*, ed. H. E. Street, pp. 329–61. London: Academic Press.

Massimino, D., André, M., Richard, C., Dagnenet, A., Massimino, J. & Vivoli, J. (1980). Évolution horaire au cours d'une journeé normale de la photosynthèse, de la transpiration, de la respiration foliaire et racinaire et de la nutrition N.P.K. chez *Zea mays*. *Physiologia Pl.* **48**, 512–18.

Masuda, H., Takahashi, T. & Sugawara, S. (1988). Acid and alkaline invertases in suspension cultures of sugar beet cells. *Plant Physiol.* **86**, 312–17.

Milthorpe, J. & Robertson, R. N. (1948). Studies in the metabolism of plant cells. 6. Salt respiration and accumulation in barley roots. *Aust. J. Exp. Biol. Med. Sci.* **26**, 189–97.

Minotti, P. L. & Jackson, W. A. (1970). Nitrate reduction in the roots and shoots of wheat seedlings. *Planta* **95**, 36–44.

Möller, I. M., Berczi, A., van der Plas, L. H. W. & Lambers, H. (1988). Measurement of the activity and capacity of the alternative pathway in intact plant tissues: identification of problems and possible solutions. *Physiologia Pl.* **72**, 642–9.

Morgan, D. R. & Street, H. E. (1959). The carbohydrate nutrition of tomato roots. 7. Sugars, sugar phosphates, and sugar alcohols as respiratory substrates for excised roots. *Ann. Bot.* **23**, 89–105.

Murray, A. J. S. & Ayres, P. G. (1986). Uptake and translocation of nitrogen in mildewed barley seedlings. *New Phytol.* **104**, 355–65.

Neales, T. F. & Davies, J. A. (1966). The effect of photoperiod duration upon the respiratory activity of the roots of wheat seedlings. *Aust. J. Biol. Sci.* **19**, 471–80.

Osman, A. M. (1971). Root respiration of wheat plants as influenced by age, temperature and irradiance of the shoots. *Photosynthetica* **5**, 107–12.

Penning de Vries, F. W. T., Witlage, J. M. & Kremer, D. (1979). Rates of respiration and of increase in structural dry matter in young wheat, ryegrass and maize plants in relation to temperature, to water stress and to their sugar content. *Ann. Bot.* **44**, 595–609.

Reynolds, J. F. & Thornley, J. H. M. (1982). A shoot:root partitioning model. *Ann. Bot.* **49**, 585–97.

Rufty, T. W., Jackson, W. A. & Raper, C. D. (1981). Nitrate reduction in roots as affected by the presence of potassium and by flux of nitrate through the roots. *Plant Physiol.* **68**, 605–9.

Saglio, P. H. & Pradet, A. (1980). Soluble sugars, respiration, and energy charge during aging of excised maize root tips. *Plant Physiol.* **66**, 516–19.

Street, H. E. (1969). Factors influencing the initiation and activity of meristems in roots. In *Root Growth*, ed. W. J. Whittington, pp. 20–38. London: Butterworths.

Theologis, A. & Laties, G. G. (1978). Relative contribution of cyto-chrome-mediated and cyanide-resistant electron transport in fresh and aged potato slices. *Plant Physiol.* **62**, 232–7.

Thornley, J. H. M. & Hesketh, J. D. (1972). Growth and respiration in cotton bolls. *J. Appl. Ecol.* **9**, 315–17.

van't Hoff, J., Hoppin, D. P. & Yagi, S. (1973). Cell arrest in G1 and G2 of the mitotic cycle of *Vicia faba* root meristems. *Am. J. Bot.* **60**, 889–95.

Visser, R. de & Blacquire, T. (1986). Inhibition and stimulation of root respiration in *Pisum* and *Plantago* by hydroxamate. Its consequences for the assessment of alternative path activity. *Plant Physiol.* **75**, 813–17.

Wagner, W., Weimken, A. & Matile, P. (1986). Regulation of fructan metabolism in leaves of barley (*Hordeum vulgare* L. cv. Gerbel). *Plant Physiol.* **81**, 444–7.

Warmbrodt, R. D. (1985). Studies on the root of *Hordeum vulgare* L. – ultrastructure of the seminal root with special reference to the phloem. *Am. J. Bot.* **72**, 414–32.

Webster, P. L. (1980). 'Stress' protein synthesis in pea root meristem cells? *Plant Sci. Lett.* **20**, 141–5.

Wetmore, R. H. & Rier, J. P. (1963). Experimental induction of vascular tissues in callus of angiosperms. *Am. J. Bot.* **50**, 418–30.

Williams, J. H. H. & Farrar, J. F. (1988). Endogenous control of photosynthesis in leaf blades of barley. *Plant Physiol. Biochem.* **26**, 503–9.

Williams, J. H. H. & Farrar, J. F. (1990). Control of barley root respira-tion by sucrose. *Physiologia Pl.* **79**, 259–66.

Willis, A. J. & Yemm, E. W. (1955). The respiration of barley plants.

VIII. Nitrogen assimilation and the respiration of the root system. *New Phytol.* **54**, 163–81.

Wright, K. & Northcote, D. H. (1972). Induced root differentiation in sycamore callus. *J. Cell Sci.* **11**, 319–37.

Yamamoto, S. & Mino, Y. (1987). Effect of sugar level on phleinase induction in stem base of orchardgrass after defoliation. *Physiologia Pl.* **69**, 456–60.

Yamamoto, S. & Mino, Y. (1989). Mechanisms of phleinase induction in the stem base of orchardgrass after defoliation. *J. Plant Physiol.* **134**, 258–60.

A. R. WELLBURN AND J. H. OWEN

Control of the rate of respiration in shoots: light, calcium and plant growth regulators

Light-grown developmental changes

Studies of the very early stages of normal light-grown plastid development in the basal intercalary meristems of cereals have shown that pronounced changes in mitochondrial activity occur before chloroplast maturation. In some respects, these events are similar to the respiratory events that precede the greening of etiolated higher plant tissues although they resemble more closely the equivalent processes in algae (see Wellburn 1982, 1984, 1987). In higher plants grown from germination onwards in natural light-dark (day-night) cycles, five distinct ultrastructural plastid stages of development have been observed – eoplasts or true proplastids, amyloplasts, amoeboid plastids (amoeboplasts), immature chloroplasts (protochloroplasts) and mature chloroplasts (Whatley, 1974, 1977; Wellburn, Robinson & Wellburn, 1982). By taking short sections from the bases of monocotyledonous seedlings, these stages may be examined separately because cells divide laterally from the mother cells in the basal intercalary meristem of young shoots and move upwards together in a regular linear manner (Esau, 1953).

Plastids, mitochondria and protoplasts may be isolated from each of these segmented regions and the interplay between organelles and bioenergetic processes studied as photosynthesis gradually predominates over respiration (see Wellburn 1982, 1984, 1987). However, the behaviour of mitochondria and the associated changes in uptake of oxygen in response to plant growth regulators and to light are interesting in their own right. This review concentrates on highlighting these latter features.

Exceptionally high rates of oxygen consumption in the dark have been detected in protoplasts isolated from segments closest to the basal intercalary meristem (Fig. 1), using either a standard oxygen electrode or a more sensitive Cartesian-diver microrespirometer (Owen, Laybourn-Parry & Wellburn, 1986). Cells in these segments only contain eoplasts

Oxygen Consumption by Protoplasts
Isolated from Barley Primary Leaf Blades

Height (cm) Upwards

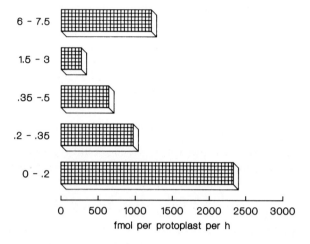

fmol per protoplast per h

Fig. 1. Oxygen consumption by protoplasts isolated from different sections of primary leaf blades of barley measuring upwards from the node of the basal intercalary meristem. Uptake of oxygen is greatest by those protoplasts isolated from sections close to the basal intercalary meristem where eoplasts (true proplastids) are in the process of conversion into amyloplasts.

and proportionately more mitochondria (by volume; Wellburn *et al.*, 1982) as well as large amounts of reducing sugars and fructans (Wellburn, Gounaris & Wellburn, 1984). These substrates are then metabolized in the cytoplasm or passed inwards to the eoplasts to form amyloplasts. Presumably, in young shoots, these reserves originally come from the seed rather than the mature part of the leaf blade but, as the seedling grows, the balance between these two sources changes.

Protoplasts from this lowermost region (0·0–0·2 cm) also display the highest proportion of cyanide-insensitive alternative pathway activity inhibited by salicylhydroxamate (SHAM). By contrast, higher segments containing protochloroplasts (1·0–1·5 cm) have the lowest rates of oxygen consumption but the highest proportion of residual (cyanide and SHAM-insensitive) oxygen uptake (Owen *et al.*, 1986).

Table 1. *Treatments responsible for significant changes in rates of oxygen exchange by segments or protoplasts isolated from the 0–0·2 cm region above the basal intercalary meristem of light-grown barley seedlings*

Strongly increase	Increase	No effect	Decrease
Blue light[a,b]	Red light[a,b]	Kinetin[b]	Far-red light[a,b]
Cycloheximide[b]	GA_3^{bd}	IAA[b]	Ca^{2+} plus ABA[c,d*]
	Uncouplers[e]	Chloramphenicol[b]	Ca^{2+} agonists[c,d]
	Paclobutrazole[e]		Ionophore A23187[d]

[a]Owen *et al.*, 1986; [b]Owen *et al.*, 1987a; [c]Owen *et al.*, 1987b; [d]Owen *et al.*, 1987c; [e]Owen & Wellburn, (unpubl.).
*Decrease is reduced by Ca^{2+} channel blockers (La^{3+}, verapamil and nifedipine) and by calmodulin antagonists (trifluoperazine and compound 40/80).

Effects of light on oxygen uptake

Light has a pronounced effect on the rates of oxygen uptake in segments close to the basal intercalary meristem (Owen, Laybourn-Parry & Wellburn, 1986, 1987a); an effect which persists in protoplasts isolated from these regions. Both blue light and red light stimulate this oxygen uptake but far-red light depresses it (Table 1). However, mitochondria isolated from these regions do not respond to light signals of any type on their own.

The far-red/red effects are due to phytochrome because classical reversion behaviour is observed but the strong blue light effect is dominant over all other photostimuli. In the field, both the coleoptiles and primary leaves of cereals act as light-pipes to lead these stimuli down to this critical area of sensitivity below soil level in the manner suggested by Mandoli & Briggs (1982, 1984).

Blue light effects on respiration and polysaccaride breakdown have long been known in algae but studies of the greening of etiolated higher plants have demonstrated that the red/far-red characteristics of phytochrome are more evident because of the parallel (or subsequent) development of plastids (Hampp & Wellburn, 1979, 1980a,b; see also Wellburn, 1984). This does not mean that blue light effects on etiolated tissues are absent. Indeed, if the light fluences are balanced, blue light enhances NADH-dependent ATP formation by mitochondria isolated from dark-grown tissues by 44% over dark controls, whereas red light only increases activity by 25% (Robinson & Wellburn, 1981). Once again, isolated mitochondria from such tissues show no responsiveness to light stimuli.

Effects of plant growth regulators on oxygen uptake

Protoplasts isolated from the meristems of seedlings treated with gibberellic acid (GA_3) show a stimulation of oxygen uptake, by comparison to that of protoplasts obtained from untreated seedlings, while abscisic acid (ABA) treatment has the opposite effect (Owen et al., 1987a; Table 1). By contrast, indole-3-acetate (IAA) and kinetin (BA) are without influence on oxygen exchange.

Only blue light treatment is able to overcome the inhibitory effects of ABA and the inhibitory effects of far-red light are removed by treatment with GA_3. In a preliminary study, no direct effects of GA_3 or ABA on NADH-dependent ATP-formation by isolated mitochondria could be detected (Robinson & Wellburn, 1981).

Effects of calcium in the presence of ABA on the consumption of oxygen by protoplasts.

Isolated meristematic protoplasts also respond directly to ABA in the incubation medium but only in the presence of calcium ions (Owen, Hetherington & Wellburn, 1987b,c; Table 1). At 10^{-3} mol m^{-3} ABA and 1 mol m^{-3} Ca^{++}, for example, the depression of oxygen uptake is 43% but this is abolished when calcium is absent or if 2 mol m^{-3} EGTA is used to chelate the cation. A range of pharmacological agents has been used to demonstrate that ABA acts as a calcium agonist in this system increasing the permeability of the plasma membrane towards calcium (Owen et al., 1987b,c; Table 1). The rise in cytoplasmic calcium then inhibits oxygen uptake by mitochondria. Antagonists of calmodulin also reduce the inhibitory effect of ABA on oxygen consumption (Owen et al., 1987b) and suggests the involvement of the cytoplasmic modulator calmodulin in the transmission of this calcium signal towards its target (Fig. 2). Recently, we repeated our earlier experiments involving isolated mitochondria (Robinson & Wellburn, 1981) using an improved method of isolation (Rickwood, Wilson & Darley-Usmar, 1987) and included measurement of changes in external calcium levels using Arsenazo III (Scarpa, 1979; Dawson, Klingenberg & Krämer, 1987) to test possible interactions between ABA and calcium on isolated organelles. Apart from verifying the observation that calcium has an effect on NAD(P)H-dependent oxygen uptake (Møller, Johnston & Palmer, 1981), we could determine no effect of ABA, either alone or in combination with calcium, on either the calcium or oxygen-exchange properties of isolated mitochondria.

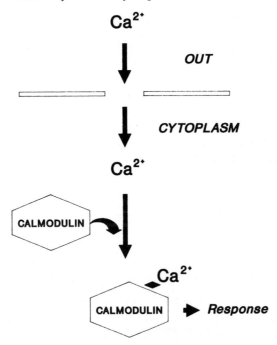

Fig. 2. Scheme to account for calcium uptake and binding to calmodulin prior to response. Amounts of free calcium in the cytoplasm are possibly also modulated by phytochrome (raised in far-red light, reduced in red) but access into the cell through the opening of calcium channels in the plasma membrane is permitted by abscisic acid (see text).

A cytoplasmic protein acts as a calcium intermediary?

In contrast to the effect of ABA, GA_3 has the capability of enhancing uptake of oxygen by isolated meristematic protoplasts but this stimulation does not require the presence of calcium ions. Interestingly, the inhibitor of cytoplasmic protein synthesis, cycloheximide, has exactly the same effect (Owen *et al.*, 1987a; Table 1). This implies that, if a particular protein or group of proteins is not produced, a sudden burst of respiratory activity takes place. In other words, a cytoplasmic product (probably protein) is either synthesized constantly, which then sets a mid-point to cellular respiratory activity, or it interacts with a calmodulin-like compound which subsequently has an effect on mitochondria, glycolysis, etc. A hypothetical model of how this might operate is shown in Fig. 3. In studies of animal tissues and S-100 proteins, a very similar mechanism for

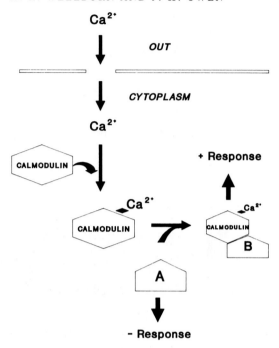

Fig. 3. Modified scheme (based on Fig. 2) to account for the free-form of a protein (A) having an inhibitory response (−) or, if modified (to B) by red light or a gibberellin-induced effect, to bind to a calcium-calmodulin complex to have exactly the opposite (+) effect.

the regulation of cellular events by calcium has already been proposed (Kligman & Hilt, 1988).

Direct effects of calcium on mitochondria

A wide range of studies have been undertaken on the uptake and efflux of calcium from mitochondria and the topic has been reviewed in this Seminar series (Earnshaw & Cooke, 1984) and more recently (Dawson *et al.*, 1987). Progress in the study of the various possible mechanisms of calcium transport into and out of plant mitochondria lag well behind equivalent studies of animal mitochondria. According to Hanson & Day (1980), plant mitochondria do not have a calcium uniporter while Russell & Wilson (1978) claim they lack potassium/calcium and sodium/calcium antiporters. Unlike vertebrate mitochondria, however, they do oxidize

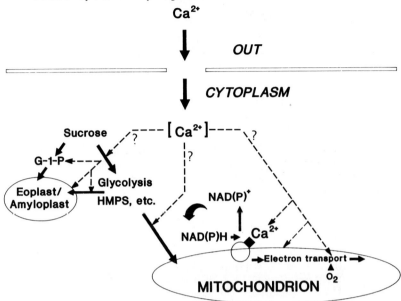

Fig. 4. Scheme to show possible effects of calcium on respiratory events. Reduced levels of free calcium are unlikely to have a direct effect on mitochondrial electron transport by means of calcium-binding to the exogenous NAD(P)H flavoprotein; although calcium uptake/efflux could be affected. However, effects of calcium on glycolysis could alter the cytoplasmic [NAD(P)H/NAD(P)$^+$] ratio which may then bring about changes in the rates of mitochondrial electron transport. Alternatively, changes in levels of free calcium could influence the apportionment of carbohydrate between glycolysis and uptake into eoplasts to form amyloplasts. The pronounced stimulatory effect of blue light upon mitochondial activity would then ensure that adequate levels of ATP are maintained throughout the cell while the known enhancement of the activity of the adenine nucleotide translocator in developing plastid envelopes (Hampp & Wellburn, 1980b) would permit sufficient ATP to enter the eoplasts and allow the formation of starch.

exogenous NAD(P)H via the respiratory chain. Møller *et al.* (1981) have shown that calcium has an effect on this process and suggest that calcium induces a change in the flavoprotein involved. Consequently, rather than calcium entry into mitochondria being primarily important for the regulation of TCA cycle dehydrogenases (Nicholls & Crompton, 1980) and pyruvate kinase (Bygrave, 1977) or as a calcium-sequestering depot for the rest of the cell, calcium may exert control on the [NAD(P)H/

NAD(P)$^+$] ratios of the plant cell at the site of exogenous NAD(P)H uptake into mitochondria. By such means, calcium influences overall respiration by exerting an effect on the rates of cytoplasmic glycolysis and oxidative pentose phosphate pathways (Earnshaw & Cooke, 1984) as well as changing the rates of mitochondrial electron transport (Fig. 4).

References

Bygrave, F. L. (1977). Mitochondrial calcium transport. In *Current Topics in Bioenergetics*, vol. 6, ed. D. R. Sanadi, pp. 259–318. New York: Academic Press.

Dawson, A., Klingenberg, M. & Krämer, R. (1987). Transport across membranes. In *Mitochondria: A Practical Approach*, ed. V. M. Darley-Usmar, D. Rickwood & M. T. Wilson, pp. 35–78. Oxford: IRL Press.

Earnshaw, M. J. & Cooke, A. (1984). The role of cations in the regulation of electron transport. In *The Physiology and Biochemistry of Plant Respiration*, ed. J. M. Palmer, SEB Seminar Series, No. 20, pp. 177–82. Cambridge: Cambridge University Press.

Esau, K. (1953). *Plant Anatomy*. pp. 79–80. New York & London: J. Wiley & Sons, Inc. & Chapman & Hall, Ltd.

Hampp, R. & Wellburn, A. R. (1979). Control of mitochondrial activities by phytochrome during greening. *Planta* **147**, 229–35.

Hampp, R. & Wellburn, A. R. (1980a). Influence of phytochrome upon mitochondrial activities during greening. In *Photoreceptors and Plant Development*, ed. J. A. DeGreef, pp. 219–28. Antwerp: Antwerpen University Press.

Hampp, R. & Wellburn, A. R. (1980b). Translocation and phosphorylation of adenine nucleotides by mitochondria and plastids during greening. *Z. Pflphysiol.* **98**, 289–303.

Hanson, J. B. & Day, D. A. (1980). Plant mitochondria. In *The Biochemistry of Plants*, vol. 1, ed. N. E. Tolbert, pp. 315–58. New York: Academic Press.

Kligman, D. & Hilt, D. C. (1988). The S100 protein family. *Trends Biochem. Sci.* **13**, 437–43.

Mandoli, D. F. & Briggs, W. R. (1982). The photoperceptive sites and the function of tissue light-piping in photomorphogenesis of etiolated oat seedlings. *Plant, Cell & Env.* **5**, 137–45.

Mandoli, D. F. & Briggs, W. R. (1984). Fiber-optic plant tissues: Spectral dependence in dark-grown and green tissues. *Photochem. Photobiol.* **39**, 419–24.

Møller, I. M., Johnston, S. P. & Palmer, J. M. (1981). A specific role for Ca^{2+} in the oxidation of exogenous NADH by Jerusalem artichoke (*Helianthus tuberosus*) mitochondria. *Biochem. J.* **194**, 487–95.

Nicholls, D. G. & Crompton, M. (1980). Mitochondrial calcium transport. *FEBS Lett.* **111**, 261–8.

Owen, J. H., Hetherington, A. M. & Wellburn, A. R. (1987b). Inhibition of respiration in protoplasts from meristematic tissues by abscisic acid in the presence of calcium ions. *J. Exp. Bot.* **38**, 498–505.

Owen, J. H., Hetherington, A. M. & Wellburn, A. R. (1987c). Calcium, calmodulin and the control of respiration in protoplasts isolated from meristematic tissues by abscisic acid. *J. Exp. Bot.* **38**, 1356–61.

Owen, J. H., Laybourn-Parry, J. E. M. & Wellburn, A. R. (1986). Leaf respiration during early plastidogenesis in light-grown barley seedlings. *Physiologia Pl.* **68**, 100–6.

Owen, J. H., Laybourn-Parry, J. E. M. & Wellburn, A. R. (1987a). Influences of light quality, growth regulations and inhibitors of protein synthesis on respiration of protoplasts from meristematic cells in barley seedlings. *Ann. Bot.* **59**, 99–105.

Rickwood, D., Wilson, M. T. & Darley-Usmar, V. M. (1987). Isolation and characteristics of intact mitochondria. In *Mitochondria: A Practical Approach*, ed. V. M. Darley-Usmar, D. Rickwood & M. T. Wilson, pp. 1–16. Oxford: IRL Press.

Robinson, D. C. & Wellburn, A. R. (1981). Light quality and hormonal influences upon the rate of ATP formation by mitochondria during greening. *Biochem. Physiol. Pfl.* **176**, 54–9.

Russell, M. J. & Wilson, S. B. (1978). Calcium transport in plant mitochondria. In *Plant Mitochondria*, ed. G. Ducet & C. Lance, pp. 175–82. Amsterdam: Elsevier.

Scarpa, A. (1979). Measurements of cation transport with metallochromic indicators. In *Methods in Enzymology*, eds. S. P. Colowick & N. O. Kaplan, vol. 61, *Biomembranes*, ed. S. Fleischer & L. Packer, pp. 301–52. New York: Academic Press.

Wellburn, A. R. (1982). Bioenergetic and ultrastructural changes associated with chloroplast development. *Intn. Rev. Cytol.* **80**, 133–91.

Wellburn, A. R. (1984). Ultrastructural respiratory and metabolic changes associated with chloroplast development. In *Topics in Photosynthesis*, vol. 5, *Chloroplast Biogenesis*, ed. N. R. Backer & J. Barber, pp. 253–303. Amsterdam: Elsevier Biomedical Press, B.V.

Wellburn, A. R. (1987). Plastids. In *Cytology and Cell Physiology*, ed. G. M. Bourne, 4th Edition, Supplement 17 to *International Review of Cytology*, pp. 149–210. New York: Academic Press.

Wellburn, A. R., Robinson, D. C. & Wellburn. F. A. M. (1982). Chloroplast development in low light-grown barley seedlings. *Planta* **154**, 259–65.

Wellburn, F. A. M., Gounaris, I. & Wellburn, A. R. (1984). Carbohydrate reserves and plant growth substance sensitivity in plastids, stomata and statocytes during shoot development. *Israel J. Bot.* **33**, 237–52.

Whatley, J. M. (1974). Chloroplast development in primary leaves of *Phaseolus vulgaris. New Phytol.* **73**, 1097–110.

Whatley, J. M. (1977). Variations in the basic pathway of chloroplast development. *New Phytol.* **78**, 407–20.

Index

Page references in *italics* refer to figures or tables.

abscisic acid (ABA), effects on oxygen uptake in shoots *191*, 192
Acer pseudoplatanus cells, autophagy triggered by sucrose deprivation 127–43
acetoacetyl-ACP synthetase, cerulenin-insensitive 26
acetyl-CoA:ACP transacylation 26, *27*
acetyl-CoA carboxylase 24–6
acyl carrier protein (ACP) 24
acyl hydrolase 139
acyl-ACP 38
  hydrolysis 31
acyl-CoA 29
acyl-CoA synthetases 29
acylated steryl glucoside 130
acyltransferases 31, 32
adenylates, control of respiration in roots 167, 171–4
ADP-glucose pyrophosphorylase 114, 116, 121
2S albumins 60–1
  biosynthesis 62–4, *64*
amino acids
  catabolized during sucrose deprivation 139–41, 143
  synthesis in roots 158–61, 182
amylopectin 111
amyloplasts 112–13
  development 190
  isolation 113
  starch breakdown following sucrose deprivation 138
  starch synthesis in 114, 115–16
amylose 111
aryloxyphenoxypropionate herbicides 24–6
asparagine, effect of sucrose starvation on 139–41, 142–3
asparagine synthetase 141

aspartate aminotransferase 141
ATP, sites of production and consumption 2
autophagic vesicles 132
autophagy, triggered by sucrose deprivation *see* sucrose, deprivation

β-oxidation 16
biogenesis
  organelle 3, 17–18
    castor bean protein bodies and glyoxysomes 59–71
    maize oil bodies 54–6
branching enzyme 114
breakage, cell 10–11
breakdown of polymers 2, 16

calcium, effects on respiration in shoots 192–6
calmodulin 192, *193*
Candida lipolytica 28
carbonylcyanide chloromethoxyphenyl hydrazone (CCCP) 171
carbonylcyanide p-trifluoromethoxyphenyl hydrazone (FCCP) 171–2
cardiolipin 130
castor bean
  allergens 60–1
  biogenesis
    of glyoxysomes 59, 65–71
    of protein bodies 59–65
  lipase 51
  organelle functions in seedling endosperm 16, *17*
  separation of endospermic organelles 13, *14*
catalase 15
  activity in maize seeds 52–4
CDP-diacylglycerol 35

cell breakage 10–11
centrifugation of organelles
    differential 11–12
    sucrose gradient 12–15
cerulenin-insensitive acetoacetyl-ACP
    synthetase 26
chloroplasts
    development 189
    starch in 111
cholinephosphotransferase 28
citrate synthase 66
compartmentation, metabolic 1, 16–18
    consequences 1–3
    demonstrated in living tissues 4–9
    enzyme localization by microscopy 15
    estimation of 5, 6
    isolation of organelles 9–15
    of nitrate assimilation in roots 149–5,
        160
Crambe 29
cyclohexanedione herbicides 24
cycloheximide, effects on oxygen uptake
    in shoots *191*, 193
cytochemistry, for enzyme localization 15
cytochrome aa_3 130

DAB (diamino benzoate) 15
DAG *see* diacylglycerol
DGDG *see* digaloctosyldiacylglycerol
diacylglycerol acyltransferase 33, 44
diacylglycerol (DAG), role in lipid
    synthesis 28, 31, 33–4, 35
diamino benzoate (DAB) 15
digalactosyldiacylglycerol (DGDG) *30*
    formation 34, *35*
dihydroxyacetone phosphate 117
dimethylsulfoxide 9
diphosphatidylglycerol 35
dithiothreitol 10
DNP 171

efflux experiments 8–9
endoplasmic reticulum
    breakage of 11
    and organelle biogenesis 18
    rough
        diacylglycerol acyltransferase in 44
        isolation by centrifugation 12, 13–14
    source of phospholipids in 16–17, 18
    and sucrose starvation 130
enolase 81, 87–8
enoyl-ACP reductase 26
enzymes
    cellular distribution 2
    localization by microscopy 15
    marker 10

erucate 29

fatty acid synthetase 26
fatty acids
    breakdown and synthesis 16
    effect of sucrose starvation on 130, *132*,
        138–9
    as respiratory substrates 130, 143
    synthesis
        de novo synthesis and the plastid 23–
            6, *27*
        the difference between '16:3' and
            '18:3'-plants *30* 30–1
        in plastids 16, 23–4, 78
        unsaturated and long chain products
            28–9, *29*
FBPase 101, 104–8
ferredoxin 157, 158, 159
firefly luciferase 69
fractionation, plant cell 9–10
    cell breakage and grinding media 10–11
    differential centrifugation 11–12
    separation on sucrose density gradients
        12–15
fructokinase 102, 113
fructose 6-phosphate 1-phosphotransferase
    enzymes *see* PFK; PFP
fructose 6-phosphate (Fru-6-P) 81, 113,
    114, 120
    interconversion to Fru-1,6-P_2 97, 102,
        104
fructose-1,6-bisphosphatase *see* FBPase
fructose-1,6-bisphosphate (Fru-1,6-P_2),
    interconversion to Fru-6-P 97, 102,
        104
fructose-2,6-bisphosphate (Fru-2,6-P_2) 82,
    85, 105
fumarase 10

α-glucan phosphorylase 138
galactolipid transferase 34
gibberellic acid (GA_3), effects on oxygen
    uptake in shoots *191*, 192, 193
11S globulin 59–60
glucokinase 113
gluconeogenesis, role of PFP in 100
glucose 1-phosphate (Glc-1-P) 102, 106,
    113, 114, 115, 116, 119
    relationship with nitrite reduction 156
glucose 6-phosphate 113
    relationship with nitrite reduction *155*,
        155–7, *157*
glucose 6-phosphate dehydrogenase 154
glucose
    compartmentation in potato slices 5
    effect on respiration in roots 182

glucose phosphate isomerase 120
glutamate synthase 141
  role in nitrate assimilation 148, 149, 150
glutamine synthetase (GS) 141, 159
  role in nitrate assimilation 148, 149,
    150–1
glyceraldehyde-3 phosphate 117
glycerol 3-phosphatase acyltransferase 31
glycerol 3-phosphate, acylation 31–2, *32*
glycerol 3-phosphate acyltransferase 27
glycolytic pathway 152
  effects of calcium on 196
  enzymes in leucoplasts 78–9, *80*, 84, 90,
    159
  role of PFK in 99–100
  role of PFP in 101
glycosides, cyanogenic 3
glycosylglycerides, formation of 33–4, *34*,
    35
glyoxysomes 59
  assembly differs from protein bodies 59,
    71
  biogenesis 66–71
  carboxy termini of enzymes 69, *70*
  structure 65–6
Golgi apparatus 2
  isolation by centrifugation 12, 13
grinding media, for tissue disintegration
    10–11

β-hydroxyacyl-ACP dehydrase 26
hexadecatrienoic acid 30
hexokinase 114, 120
hexose monophosphate
  from starch breakdown 138
  interconversion to triose phosphate 95–
    108
  relationship with nitrite reduction 154–5
  see also glucose 1-phosphate; glucose 6-
    phosphate

immunoelectron microscopy 15
invertase 113, 181
isocitrate lyase 67, *68*, 139
  in maize seeds 52–4
isozymes, in leucoplasts *see* leucoplasts

β-ketoacyl-ACP reductase 26
β-ketoacyl-ACP synthetase 2, 26
ketose reductase 113
Krebs cycle 152
  see also TCA cycle

7S lectins 60

leucoplasts
  in the castor plant endosperm 78–9, *80*
  isozymes in 81, 90–1
    ATP-dependent phosphofructokinase
      (PFK) 82–4, 98, 99
    enolase 87–8
    phosphoglyceromutase 86–7
    pyrophosphate-dependent
      phosphofructokinase (PFP), 85–6,
      100–4
    pyruvate kinase 88–90
  see also amyloplasts
light, effects on oxygen uptake in shoots
    191
linoleate 32
  desaturation 28
linoleic acid 26
linolenate synthesis 28, *29*
6-linolenic acid 26
lipase
  activity in seeds 51, 57
    of maize 52–4, 55, *56*
  biosynthesis in maize 52, 54–5
  localized in maize oil bodies 51–2
lipid bodies *see* oil bodies
lipid exchange proteins, possible roles 36–
    8
lipid synthesis 23
  acylation of glycerol 3-phosphate 31–2,
    32
  differences in '16:3' and '18:3'-plants *37*,
    38
  fatty acids *see* fatty acids, synthesis
  glycosylglyceride formation 33–4, *35*
  phosphoglyceride synthesis 35–6, *36*
  possible roles of lipid exchange proteins
    36–8
  triacylglycerol accumulation 33
lipids, movement within the cell 36–8
luciferase, firefly 69
lysophosphatidate acyltransferase 30–1, 32
lysophosphatidylcholine:acetyl-CoA
    acyltransferase 28

malate
  demonstration of different pools in corn
    roots 7–8, *8*
  different rates of turnover 4
  metabolism linked to nitrate reduction
    152–4
  in vacuoles 4
malate dehydrogenase, glyoxysomal 69
malate synthase 66, 67, 69, 139
malonate 153–4
malonyl-CoA 25, 29
membranes, ion gradients 2

metabolites
  compartmentation within cells 3–9
  transport within cells 3
  see also compartmentation, metabolic
MGDG (monogalactosyldiacylglycerol)
    28, *30*, 31, 34
microbodies (peroxisomes) 65
  glyoxysomes see glyoxysomes
microscopy, for enzyme localization 15
microsomal pellets
  from centrifugation down a sucrose
      gradient 13–14
  from differential centrifugation 12
mitochondria
  centrifugal 'mitochondrial pellets' 11
  effects of calcium on 194–6
  effects of sucrose starvation on 129–30
  enzyme concentrations 2
  marker enzyme for intact 10
mitochondrial electron transport chain
    (METC) 172–3
  effects of sucrose on 182
monogalactosyldiacylglycerol (MGDG)
    28, *30*, 31, 34
NADH, for nitrate reductase 152–4
nitrate reductase (NR) 148–9, *150*, 182
  origin of reductant in roots 151–4, 159
nitrates
  assimilation in roots 147–8, *160*
  compartmentation of 149–51, 160
  integration of nitrogen and carbon
      metabolism 151–4, 159
  origin of reductant for nitrite
      reductase 154–8
  pathway 148–9
  and respiration rate 174, 175, *176*
  assimilation in shoots 147–8
nitrite reductase 148, 151
  origin of reductant in roots 154–8
NMR techniques
  13C-NMR 8
    investigation of starch synthesis 117
  31P-NMR 8
    to study the effects of sucrose
      starvation 132–9
NR see nitrate reductase
nuclear magnetic resonance techniques see
    NMR techniques

oil bodies 43–4
  function of oleosins 56–7
  lipase activity in 51–4
  membrane of 43–51, 56–7
  synthesis and degradation in maize 54–
      6, *56*
  see also oleosins

oil seed rape, developing acetyl-CoA
    carboxylase 24, *25*
oleic acid 26
oleosins 44–5, *45*
  in diverse seed species 49–50
  function 56–7
  genetic encoding 50–1
  structural features *47*, 47–9, 50
  synthesized in the rough ER 45–6, *46*, 54
oleosomes see oil bodies
oleoyl-ACP 27
oleoyl-ACP thioesterase 27
oligomycin 129
OPPP (oxidative pentose phosphate
    pathway) 152, 154–6, 196
organelles
  biogenesis see biogenesis, organelle
  consequences of compartmentation in
      1–3
  functions in castor bean seedling
      endosperm 16, *17*
  involved in oxygen assimilation in roots
      159–61
  isolation 9–15
  marker enzymes for 10
  see also endoplasmic reticulum;
      glyoxysomes; mitochondria; oil
      bodies; plastids; protein bodies
oxaloacetate 141
oxidative pentose phosphate pathway
    (OPPP) 118, 152, 154–6, 196
2-oxoglutarate 159
oxygen uptake see respiration

palmitate 30, 31, 32
  enzyme complexes used in formation of
      24–6
palmitoyl-ACT 26, 30, 31
palmitoyl-CoA 31
PEP (phosphoenol pyruvate) 79
peroxisomes 65
  glyoxysomes see glyoxysomes
PFK (phosphofructokinase) 81–2
  ATP-dependent 82–4, 95, 98–100, 121
  pyrophosphate-dependent see PFP
PFP (pyrophosphate dependent fructose-
      6-phosphate 1-phosphotransferase)
      81–2, 85–6, 95, 100–4, 121
PGM (phosphoglyceromutase) 79, 81, 86–
      7
pH, in different areas of the cell 2
phleinase 181
phosphatases, localization by microscopy
      15
phosphatidate 31–2, *32*
  dephosphorylation 33

phosphatidate phosphohydrolase 33, 34
phosphatidic acid 32
phosphatidylcholine (PC) 28, 35–6, 142
phosphatidylethanolamine (PE) 35–6
phosphatidylglycerol (PG) 35
phosphatidylinositol (PI) 35–6
phosphoenol pyruvate (PEP) 79, 98
phosphofructokinase *see* PFK
phosphoglucomutase 120
6-phosphogluconate dehydrogenase 154
2-phosphoglycerate 79
phosphoglycerides
    distribution 35
    synthesis 35–6, *36*
phosphoglyceromutase (PGM) 79, 81, 86–7
phospholipid synthesis 17, 18
phosphorylcholine 139, 142
photosynthesis, control by sucrose 180
pigments, in vacuoles 3
PK *see* pyruvate kinase
plasmalemma, breakage of 11
plastids 77
    isolation by centrifugation 13
    matrix pH 2
    properties 77–8
    site of fatty acid synthesis 16, 23–4, 78
    stages of development 189
    see also amyloplasts; chloroplasts;
        leucoplasts
polyvinyl pyrolidone 10
potatoes, compartmentation of glucose in
    5
PP_i (inorganic pyrophosphate)
    possible role of PFP in synthesis of 101–
        4
    regulation by PFP 103–4, 108
    roles 102–3, 103, 116
preproalbumin 62, *63*, *64*
preproricin 61–2, *63*
proricin 62
protein bodies 59
    assembly differs from glyoxysomes 59,
        71
    biogenesis 61–5
    structure 59–61
proteins
    breakdown, during sucrose starvation
        139–41, 143
    synthesis in roots 158–61
        effect of sugar 182
protoplasts, isolation 14–15
    respiration 189–92
pruning, effect on root metabolism 176–9,
    177, *178*, *179*
pyridine nucleotide reductase 157–8, 159

pyrophosphatase 103, 116
pyrophosphate dependent fructose-6-
        phosphate 1-phosphotransferase *see*
        PFP
pyrophosphate *see* PP_i
pyrophosphorylases 103, 116
pyruvate, supply to plastids 78–9
pyruvate dehydrogenase complex 78
pyruvate kinase (PK) 81, 88–90
    leucoplast 79

rape *see* oil seed rape
RCA (*Ricinus communis* agglutinin) 60
respiration
    effects of sucrose starvation 127–30
    relationship with growth 174–5, *175*, *176*
    in roots 167
        capacity affected by sucrose supply
            176–9
        compartmentation and fluxes of
            carbohydrate 167–70
        control by adenylates 171–4
        control by substrates 170–1
        effect of pruning on 176–9, *177*, *178*,
            179
        functions 174–5, *175*, *176*
        sucrose providing coarse control 180–
            3
    in shoots
        effects of calcium 192–6
        effects of light on oxygen uptake *191*
        effects of plant growth regulators 192,
            193
        light-grown developmental changes
            189–90, *190*
ribosomes, cytoplasmic 18
ricin 3
    biosynthesis 61–2, *63*, *64*
    in castor bean seeds 60
Ricinus communis agglutinin (RCA) 60
Ricinus communis see castor bean
roots
    carbon and nitrogen metabolism in 147–
        61
    compartmentation and fluxes of
        carbohydrate 167–70
    pools of malate in corn roots 7–8, *8*
    respiration *see* respiration, in roots

salicylhydroxamate (SHAM) 173, 190
seeds
    castor bean *see* castor bean
    post-germinative growth 51, 66
    starch biosynthesis in wheat 111–21
    storage oils 43–57
Selanastrum minutum 89

senescence 127
SHAM (salicylhydroxamate) 173, 190
spherosomes *see* oil bodies
starch
  breakdown 96, 138
  importance 111
  interconversion to sucrose 95, 96, 100
  synthesis 95, 102, 104, 111–13, 119–21,
    120, 121
    early steps 114–16
    enzymes involved in 113, 119–20, *120*
    evaluation of triose phosphate
      involvement 106, 107, 116–19, *118,
      119*
    last steps 114
    in storage tissues 112–13
    'textbook' pathway *112*
starch synthase 114, 121
stearate 26
stearoyl-ACP 26, 27
stearoyl-ACP desaturase 27
steryl glucoside 130
sucrose
  breakdown 102–3
  compartmentation in roots 168–70
  deprivation 127
    effect on asparginine accumulation
      139–41
    effect on oxygen consumption 127–30
    effect on polar lipid fatty acids 130,
      131, 132, *132*
    effect of replenishment 142–3
    31P-NMR of cells 132–9
  flux into roots 168
  interconversion to starch 95–6
  and respiration in roots 167, 176–83
  role of PFP in synthesis 100–1
  see also respiration, in roots
sucrose phosphatase 120
sucrose phosphate synthase 120
sucrose synthase 102, 113, 114, 119, 120,
  181

sucrose-sucrose fructosyl transferase 181
sycamore cells, autophagy triggered by
  sucrose deprivation 127–43
synthesis, and breakdown of polymers 2,
  16

TCA cycle, estimating compartmentation
  in *6, 7*
  see also Krebs cycle
tonoplast, breakage of 11
toxic materials, importance of
  compartmentation 3
transport
  of metabolites 3, 16
  through the plasmalemma 2
triacylglycerols, accumulation 33, 43
tricarboxylic acid cycle *see* TCA cycle
triose phosphates
  from starch breakdown 138
  interconversion to hexose
    monophosphates 95–108
  involved in starch synthesis 112–13,
    114–15, 116–19
triose-phosphate isomerase 117

ubiquinone 172
ubiquitin 130
UDP-glucose 102, 116, 119, 127
UDP-glucose pyrophosphorylase 102, 113,
  114, 116, 119
uncouplers 171–2

vacuoles
  compartmentation of products in 3
  malate in 4
  pH 2

waste products, importance of
  compartmentation 3
wheat grains, starch synthesis 111–21